CAMBRIDGE TRACTS IN MATHEMATICS

General Editors
B. BOLLOBAS, H. HALBERSTAM & C. T. C. WALL

---

97 Algebraic curves over finite fields

T0282313

In this Tract, Professor Moreno develops the theory of algebraic curves over finite fields, their zeta and L-functions, and, for the first time, the theory of algebraic geometric Goppa codes on algebraic curves.

Amongst the applications considered are: the problem of counting the number of solutions of equations over finite fields; Bombieri's proof of the Riemann hypothesis for function fields, with consequences for the estimation of exponential sums in one variable; Goppa's theory of error-correcting codes constructed from linear systems on algebraic curves; there is also a new proof of the Tsfasman–Vladut–Zink theorem.

The prerequisites needed to follow this book are few, and it can be used for graduate courses for mathematics students. Electrical engineers who need to understand the modern developments in the theory of error-correcting codes will also benefit from studying this work.

CARLOS MORENO

Professor of Mathematics
Baruch College
City University of New York

Algebraic curves over finite fields

CAMBRIDGE
UNIVERSITY PRESS

CAMBRIDGE UNIVERSITY PRESS
Cambridge, New York, Melbourne, Madrid, Cape Town, Singapore,
São Paulo, Delhi, Dubai, Tokyo, Mexico City

Cambridge University Press
The Edinburgh Building, Cambridge CB2 8RU, UK

Published in the United States of America by
Cambridge University Press, New York

www.cambridge.org
Information on this title: www.cambridge.org/9780521459013

First published 1991
First paperback edition 1993

A catalogue record for this publication is available from the British Library

Library of Congress cataloguing in publication data available

ISBN 978-0-521-34252-0 Hardback
ISBN 978-0-521-45901-3 Paperback

Contents

Contents

Preface

This is an introduction to the theory of algebraic curves over finite fields. There are three main themes. The first is a complete presentation of Bombieri's proof of the Riemann hypothesis in the function field case. The second is a full development of the theory of exponential sums in one variable from the point of view of Hasse and Weil. The third and most novel part is the theory of error correcting codes following the program outlined by Goppa. The new results in this last area have come to depend increasingly on many ideas from the theory of modular curves over finite fields and to some extent have motivated our overall presentation. We have included two introductory chapters, one on the basic notions about algebraic curves and the associated function fields, the other including a proof of the Riemann–Roch theorem. In an appendix we verify constructively how the singularities of a plane algebraic curve defined by a homogeneous polynomial in three variables can be transformed into ordinary singularities, i.e. points with distinct tangents, at the expense of increasing the field of constants. This is an essential step in Goppa's program of constructing error correcting codes on algebraic curves from their linear systems. This book fills a gap in the literature of modern number theory; it makes available for the first time all the known results about exponential sums which have applications in algebraic and analytic number theory. Chapter 5 on error correcting codes and the appendix may be studied independently of the rest of the book; they are intended mostly for workers in the field who want to understand the new results about codes on algebraic curves over finite fields.

1

Algebraic curves and function fields

1.1 Geometric aspects

1.1.1 Introduction

In applications to arithmetical questions and coding theory, the basic field of constants will be the finite field $k = \mathbb{F}_0$ of characteristic p; in particular this will be apparent in the proof of the Riemann–Roch theorem as well as in the study of the zeta function of a curve. In the present chapter however the finite field $\mathbb{F}_0$ is replaced by its algebraic closure $\mathbb{F}$ which we can think of as the union of all finite extensions of $\mathbb{F}_0$. As a consequence of this choice of base field, the study of most of the geometric properties of curves and their connection with the algebraic properties of function fields can be guided by our geometric intuition as if we were working over the field of complex numbers.

In Section 1.1 we present the definitions necessary for an understanding of the fundamental properties of algebraic curves. The main result is Theorem 1.1 in Section 1.1.7 which establishes the existence of an algebraic curve which is a smooth model associated to a given function field K/k. We give there only a summary of the key results; the interested reader desiring more details is advised to consult the first chapter of Hartshorne's book [34] or Chapter II, Section 5 of Shafarevitch's book [76]. Section 1.2 is of a more technical nature and deals exclusively with the algebraic properties of the valuation rings of the function field of a curve. In a sense this part is a preparation for the proof of the Riemann–Roch theorem given in Chapter 2. The reader can find a more complete treatment in Chevalley ([11], Ch. I).

1.1.2 Affine varieties

Affine n-space over k is the set $\mathbb{A}^n$ consisting of all n-tuples of elements of k. If p is a point in $\mathbb{A}^n$ and $p = (a_1, \ldots, a_n)$, then the $a_i \in k$ are called the coordinates of p. $A = k[x_1, \ldots, x_n]$ denotes the polynomial ring in n variables over k. The evaluation of a polynomial $f \in A$ at points in $\mathbb{A}^n$ gives a map $f: \mathbb{A}^n \to k$. The set of points $p \in \mathbb{A}^n$ where $f(p) = 0$ is called the set of zeros of f. More generally if $T \subset A$, we associate with T the subset of $\mathbb{A}^n$ given by

$$Z(T) = \{p \in \mathbb{A}: f(p) = 0 \text{ for every } f \in T\}.$$

In particular if $\mathscr{S}$ is the ideal of A generated by T, we put $Z(T) = Z(\mathscr{S})$. A fundamental property of the ring A is its noetherian nature, which implies that any ideal $\mathscr{S}$ has a finite set of generators, i.e. $\mathscr{S} = (f_1, \ldots, f_r)$.

Definition 1.1 *A subset Y of $\mathbb{A}^n$ is called an algebraic subset if there exists a subset $T \subset A$ such that $Y = Z(T)$.*

Definition 1.2 *The Zariski topology on $\mathbb{A}^n$ is defined by taking the open subsets to be the complements of the algebraic sets.*

To see why this indeed defines a topology one needs to verify that the algebraic sets, which in this topology play the role of closed sets, satisfy the following properties: (i) the union of two algebraic sets is an algebraic set, (ii) the intersection of any family of algebraic sets is an algebraic set, (iii) the empty set and the whole space are algebraic sets. Only the second requirement is not obvious; it follows from the noetherian nature of A.

Example The affine line

$$\mathbb{A}^1 \quad \underset{a \qquad b}{\rule{4cm}{0.4pt}}$$

Since $A = k[x]$ is a principal ideal domain, any ideal in A is of the form $(p(x))$, where $p(x)$ is a polynomial with coefficients in k. Since k is assumed to be algebraically closed, $p(x)$ splits into a finite number of linear factors. Hence the algebraic set associated with the ideal $(p(x))$ consists of the roots of $p(x)$. Therefore the closed sets of $\mathbb{A}^1$ are $\mathbb{A}^1$ itself and all finite subsets. Observe that any two open subsets in $\mathbb{A}^1$ always have a non-void intersection.

A non-empty subset Y of a topological space X is said to be *irreducible* if it cannot be expressed as the union $Y = Y_1 \cup Y_2$ of two proper subsets, each of which is closed in Y. The empty set is not considered to be irreducible.

Definition 1.3 *An affine variety is an irreducible closed subset of $\mathbb{A}^n$. An open subset of an affine variety is a quasi-affine variety.*

If Y is a subset of $\mathbb{A}^n$, we define the ideal of Y in $A = k[x_1, \ldots, x_n]$ by

$$I(Y) = \{f \in A : f(p) = 0 \text{ for all } p \in Y\}.$$

The Hilbert Nullstellensatz guarantees that the maps $Y \to I(Y)$ and $\mathscr{S} \to Z(\mathscr{S})$ defined above set up a one-to-one correspondence between algebraic sets in $\mathbb{A}^n$ and ideals $\mathscr{S}$ in A which satisfy

$$\mathscr{S} = \{f \in A : f^r \in \mathscr{S} \text{ for some integer } r > 0\},$$

i.e. $\mathscr{S}$ is its own radical. Under this correspondence an irreducible algebraic set in $\mathbb{A}^n$ is associated to a prime ideal.

Example Let f be an irreducible polynomial in $A = k[x, y]$. Since A is a unique factorization domain, f generates a prime ideal (f) and the corresponding algebraic set $Y = Z(f)$ is irreducible. If f is of degree d, then Y is said to be a curve of degree d.

A maximal ideal m of $A = k[x_1, \ldots, x_n]$ corresponds to a minimal irreducible closed subset of $\mathbb{A}^n$; in fact $Z(m)$ must be a point, say $p = (a_1, \ldots, a_n)$. Therefore every maximal ideal of A is of the form $m = (x_1 - a_1, \ldots, x_n - a_n)$ for some $a_1, \ldots, a_n$ in k.

Definition 1.4 *If* $Y \subseteq \mathbb{A}^n$ *is an affine algebraic set, we define the affine coordinate ring* $A(Y)$ *to be the quotient* $A/I(Y)$.

Remark If Y is an affine variety, then $I(Y)$ is a prime ideal and therefore $A(Y)$ is an integral domain. The coordinate ring $A(Y)$ of an affine algebraic set is a finitely generated k-algebra. Conversely, any finitely generated k-algebra which is a domain is the affine coordinate ring of some affine variety; in fact such a k-algebra is the quotient of a polynomial ring $A = k[x_1, \ldots, x_n]$ by an ideal $\mathscr{S}$, in which case we can take as affine variety $Y = Z(\mathscr{S})$.

A topological space Y is called noetherian if for any sequence $Y_1 \supseteq Y_2 \supseteq \cdots$ of closed subsets there is an integer r such that $Y_r = Y_{r+1} = \cdots$. A basic property of a noetherian space X is that every non-empty closed subset Y can be expressed uniquely as a finite union $Y = Y_1 \cup \cdots \cup Y_r$ where the Y_i are irreducible sets and $Y_i \not\supseteq Y_j$ for $i \neq j$. The Y_i are called the irreducible components of Y. Since $\mathbb{A}^n$ is a noetherian space, a property inherited from the polynomial ring $A = k[x_1, \ldots, x_n]$, we see that every algebraic set can be represented uniquely as a union of varieties, no one containing another.

Definition 1.5 *If X is a topological space, we define the dimension of X to be the supremum of all integers n such that there exists a chain $Z_0 \subset Z_1 \subset \cdots \subset Z_n$ of distinct irreducible closed subsets of X. The dimension of an affine or quasi-affine variety is defined to be its dimension as a topological space.*

Example The dimension of the affine line $\mathbb{A}^1$ is clearly 1, since the only irreducible closed subsets of $\mathbb{A}^1$ are the whole space and the single points.

The dimension of an affine variety Y is also equal to the Krull dimension of its affine coordinate ring $A(Y)$. If k is a field, and B is a domain, which is also a finitely generated k-algebra, then the Krull dimension of B is equal to the transcendence degree of the quotient field $K(B)$ of B over k. For example dim $\mathbb{A}^n = n$ since $K(\mathbb{A}^n) = k(x_1, \ldots, x_n)$ is of transcendence degree n over k.

Remark If Y is a quasi-affine variety, then dim $Y = \dim \bar{Y}$, where $\bar{Y}$ is the closure of Y in the Zariski topology.

Since the noetherian integral domain $A = k[x_1, \ldots, x_n]$ is a unique factorization domain, a variety Y in $\mathbb{A}^n$ has dimension $n - 1$ if and only if it is the zero set $Z(f)$ of a single non-constant irreducible polynomial in A. Thus the algebraic curves in $\mathbb{A}^2$ are in one-to-one correspondence with the non-constant irreducible polynomials in $A = k[x, y]$.

1.1.3 Projective varieties

Projective n-space over k is the set $\mathbb{P}^n$ consisting of all equivalence classes of $(n + 1)$-tuples $(a_0, \ldots, a_n)$ of elements of k, not all zero, under the equivalence relation given by $(a_0, \ldots, a_n) \sim (\lambda a_0, \ldots, \lambda a_n)$ for all $\lambda \in k$, $\lambda \neq 0$. As a set $\mathbb{P}^n$ can be identified as the space of all lines in $\mathbb{A}^{n+1}$ passing through the origin. An element in $\mathbb{P}^n$ is called a point. If $p \in \mathbb{P}^n$ is a point, then any $(n + 1)$-tuple $(a_0, \ldots, a_n)$ in the equivalence class of p is called a *set of homogeneous coordinates* for p. On the polynomial ring $S = k[x_0, \ldots, x_n]$ we introduce a grading

$$S = \bigoplus_{d \geq 0} S_d,$$

where S_d is the set of all linear combinations of monomials of total weight d in $x_0, \ldots, x_n$. Clearly we have for any $d, e \geq 0$ that $S_d S_e \subseteq S_{d+e}$. An element of S_d is called a homogeneous polynomial of degree d. It is obvious that any element of S has a unique decomposition as a finite sum of homogeneous polynomials. An ideal $\mathscr{S} \subseteq S$ is called a homogeneous ideal if $\mathscr{S} = \bigoplus_{d \geq 0}(\mathscr{S} \cap S_d)$. It can be shown that an ideal is homogeneous if and only if it can be generated by homogeneous elements. The sum, product, intersection and radical of homogeneous ideals are homogeneous. To verify that a homogeneous ideal $\mathscr{S}$ in S is prime, it suffices to show for any two homogeneous elements f, g that $fg \in S$ implies $f \in \mathscr{S}$ or $g \in \mathscr{S}$.

If $f \in S_d$, i.e. $f(\lambda x_0, \ldots, \lambda x_n) = \lambda^d f(x_0, \ldots, x_n)$, then we have a well-defined map $f: \mathbb{P}^n \to \{0, 1\}$, where $f(p) = 0$ if $f(a_0, \ldots, a_n) = 0$ and $f(p) = 1$ if $f(a_0, \ldots, a_n) \neq 0$, where $(a_0, \ldots, a_n)$ is a set of homogeneous coordinates for p. If f is a homogeneous polynomial, we define the set of zeros of f to be the set of points p in $\mathbb{P}^n$ where $f(p) = 0$. If T is a set of homogeneous elements of S, we define the zero set of T to be

$$Z(T) = \{p \in \mathbb{P}^n : f(p) = 0 \text{ for all } f \in T\}.$$

If $\mathscr{S}$ is a homogeneous ideal in S, we define $Z(\mathscr{S}) = Z(T)$, where T is the set of all homogeneous elements in $\mathscr{S}$. Since S is a noetherian ring, we have as in the affine case, that any set T of homogeneous elements has a finite subset $f_1, \ldots, f_r$ such that $Z(T) = Z(f_1, \ldots, f_r)$.

Definition 1.6 *A subset Y of $\mathbb{P}^n$ is an algebraic set if there exists a set T of homogeneous elements of S such that $Y = Z(T)$.*

Definition 1.7 *The Zariski topology on $\mathbb{P}^n$ is defined by taking the open sets to be the complements of algebraic sets.*

Again to verify that indeed this definition makes sense one checks that: (i) the union of two algebraic sets is an algebraic set, (ii) the intersection of any family of algebraic sets is an algebraic set and (iii) the empty set and the whole space are algebraic sets.

Definition 1.8 *A projective variety is an irreducible algebraic set in $\mathbb{P}^n$, endowed with the induced topology. An open subset of a projective variety is called a quasi-projective variety.*

The dimension of a projective or quasi-projective variety is its dimension as a topological space. Let Y be any subset of $\mathbb{P}^n$; the homogeneous ideal of Y in S is the ideal $I(Y)$ generated by the set

$$\{f \in S : f \text{ is homogeneous and } f(p) = 0 \text{ for all } p \in Y\}.$$

If Y is an algebraic set, we define the homogeneous coordinate ring of Y to be $S(Y) = S/I(Y)$.

An important property of projective n-space $\mathbb{P}^n$ is that it has an open covering consisting of affine n-spaces. Let us see how this comes about. If f is a linear homogeneous polynomial then the zero set of f is called a *hyperplane*. In particular we denote the zero set of x_i by H_i for $i = 0, \ldots, n$. Let U_i be the open set $\mathbb{P}^n - H_i$. If p is any point in $\mathbb{P}^n$ with homogeneous coordinates $(a_0, \ldots, a_n)$, then $a_i \neq 0$ for at least one i and hence $p \in H_i$. This shows that $\mathbb{P}^n$ is covered by the open sets U_i. We now define a map

$$\varphi_i \colon U_i \to \mathbb{A}^n$$

as follows: if $p \in U_i$ has homogeneous coordinates $(a_0, \ldots, a_n)$, then $\varphi_i(p) = q$, where q is the point in $\mathbb{A}^n$ with affine coordinates

$$(a_0/a_i, \ldots, a_n/a_i),$$

with a_i/a_i omitted. The map φ_i is well defined since the ratios a_j/a_i are independent of the choice of homogeneous coordinates. The map φ_i is in fact a homeomorphism from U_i with its induced topology to $\mathbb{A}^n$ with its Zariski topology. This property of $\mathbb{P}^n$ is also inherited by projective and quasi-projective varieties: if Y is a projective (respectively, quasi-projective) variety, then Y is covered by the open sets $Y \cap U_i$, $i = 0, \ldots, n$ which are homomorphic to affine (respectively, quasi-affine) varieties via the map φ_i.

1.1.4 Morphisms

Let Y be a quasi-affine variety in $\mathbb{A}^n$.

Definition 1.9 *A function $f \colon Y \to k$ is regular at a point $p \in Y$ if there is an open neighborhood U with $p \in U \subseteq Y$, and polynomials $g, h \in k[x_1, \ldots, x_n]$ such that h is nowhere zero on U, and $f = g/h$ on U. We say f is regular on Y if it is regular at every point of Y.*

Remark If k is identified with the affine line $\mathbb{A}^1$ together with its Zariski topology, then a regular function $f \colon Y \to \mathbb{A}^1$ is continuous.

Let $Y \subseteq \mathbb{P}^n$ be a quasi-projective variety.

Definition 1.10 *A function $f \colon Y \to k$ is regular at a point $p \in Y$ if there is an open neighborhood U with $p \in U \subseteq Y$, and homogeneous polynomials $g, h \in S = k[x_0, \ldots, x_n]$ of the same degree, such that h is nowhere zero on U and $f = g/h$ on U. We say that f is regular on Y if it is regular at every point.*

In the following an affine, a quasi-affine, a projective or a quasi-projective variety will simply be referred to as a variety.

Definition 1.11 *If X, Y are two varieties, a morphism*

$$\varphi \colon X \to Y$$

is a continuous map such that for every open set $V \subseteq Y$, and for every regular function $f \colon V \to k$, the function $f \circ \varphi \colon \varphi^{-1}(V) \to k$ is regular.

Let Y be a variety. We denote by $\mathcal{O}(Y)$ the ring of all regular functions on Y. If p is a point of Y, we define the local ring of p on Y, $\mathcal{O}_{p,Y}$ (or simply $\mathcal{O}_p$) to be the local ring of germs of regular functions on Y near p. Recall that a germ of a regular function at p is the equivalence class of pairs $\langle U, f \rangle$, where U is an open subset of Y containing p, and f is a regular function on U, and where two such pairs $\langle U, f \rangle$ and $\langle V, g \rangle$ are said to be equivalent if $f = g$ on $U \cap V$. $\mathcal{O}_p$ is a local ring and its maximal ideal m is the set of germs of regular functions which vanish at p. Since we are assuming that the base field k is algebraically closed, we have that the residue field $k_p = \mathcal{O}_p / m$ is isomorphic to k.

Definition 1.12 *The function field $K(Y)$ of a variety Y is the set of all equivalence classes of pairs $\langle U, f \rangle$ where U is a non-empty subset of Y, f is a regular function on U, and where we identify two pairs $\langle U, f \rangle$ and $\langle V, g \rangle$ if $f = g$ on $U \cap V$. The elements of $K(Y)$ are called the rational functions on Y.*

The natural maps $\mathcal{O}(Y) \to \mathcal{O}_p \to K(Y)$ are clearly injective. If $Y \subseteq \mathbb{A}^n$ is an affine algebraic variety, then its affine coordinate ring $A(Y)$ is isomorphic to the ring of all regular functions $\mathcal{O}(Y)$. If to each point $p \in Y$ we associate the ideal $m_p \subseteq A(Y)$ of functions vanishing at p, then the map $p \to m_p$ sets up a one-to-one correspondence between the points of Y and the maximal ideals of $A(Y)$. Furthermore, for each point p, the localization $A(Y)_{m_p}$ of $A(Y)$ at the ideal m_p is isomorphic to the local ring $\mathcal{O}_p$. The quotient field of the affine coordinate ring $A(Y)$ is isomorphic to the function field $K(Y)$, and hence it is a finitely generated extension field of k of transcendence degree $= \dim Y$.

Consider $S = k[x_0, \ldots, x_n]$ as a graded ring and let $\mathcal{Q}$ be a homogeneous prime ideal in S, let T be the multiplicative subset of S consisting of the homogeneous elements not in $\mathcal{Q}$. The localization $S_{\mathcal{Q}} = T^{-1}S$ of S with respect to T has a natural grading given by $\deg(f/g) = \deg f - \deg g$ for f a homogeneous element on S and $g \in T$. The subring

$$S_{(\mathscr{D})} = \{f/g \in T^{-1}S : \deg(f/g) = 0\}$$

is a local ring with maximal ideal $(\mathscr{D} \cdot T^{-1}S) \cap S_{(\mathscr{D})}$. If S is a domain and $\mathscr{D} = (0)$, then $S_{(0)}$ is a field. If $f \in S$ is a homogeneous element, we denote by $S_{(f)}$ the subring of elements of degree 0 in the localized ring S_f.

Let $S(Y)$ be the homogeneous coordinate ring of a projective variety $Y \subseteq \mathbb{P}^n$. A characteristic property of such varieties is that the ring of all regular functions is identical to the field of constants:

$$\mathcal{O}(Y) = k,$$

i.e. the only functions which are regular everywhere on a projective variety are the constants. Let p be any point on Y and define the ideal

$$m_p = \{f \in S(Y) : f \text{ homogeneous and } f(p) = 0\}.$$

We then have $\mathcal{O}_p = S(Y)_{(m_p)}$. Furthermore if $K(Y)$ is the function field of Y, then $K(Y) \simeq S(Y)_{((0))}$.

Remark The correspondence $X \to A(X)$ which associates to a variety its coordinate ring and the analogous correspondence that sends a point $p \in X$ to its local ring $\mathcal{O}_p$ are the key to the reduction of geometric problems to questions about finitely generated k-algebras.

Finally we mention the well-known result concerning the finiteness of integral closure, a result that will be useful later on. Let A be an integral domain which is a finitely generated algebra over a field k. Let K be the quotient field of A, and let L be a finite algebraic extension of K. Then the integral closure A' of A in L is a finitely generated A-module, and is also a finitely generated k-algebra.

1.1.5 Rational maps

Let φ and ψ be two morphisms from the variety X to the variety Y. Suppose φ and ψ agree when restricted to a non-empty open subset. The image $(\varphi \times \psi)(U)$ of U under the product map $\varphi \times \psi : X \to Y \times Y$ $(\varphi \times \psi)(x) = (\varphi(x), \psi(x))$ is a dense subset of the diagonal $\Delta_Y = \{(y, y) : y \in Y\}$, which is itself a closed subset. The closure $(\varphi \times \psi)(X)$ is therefore also contained in Δ_Y. This shows that φ and ψ agree everywhere.

Definition 1.13 *Let X and Y be varieties. A rational map $\varphi : X \to Y$ is an equivalence class of pairs $\langle U, \varphi_U \rangle$, where U is a non-empty open subset of*

X, φ_U is a morphism of U to Y, and where $\langle U, \varphi_U \rangle$ and $\langle V, \varphi_V \rangle$ are equivalent if φ_U and φ_V agree on $U \cap V$. The rational map φ is dominant if for some pair $\langle U, \varphi_U \rangle$, the image of φ_U is dense in Y.

Definition 1.14 *Two varieties X and Y are called birationally isomorphic, or simply isomorphic if there is a rational map $\varphi \colon X \to Y$ which admits an inverse, namely a rational map $\psi \colon Y \to X$ such that $\psi \circ \varphi = id_X$ and $\varphi \circ \psi = id_Y$ as rational maps.*

Let $Y \subseteq \mathbb{A}^n$ be a hypersurface in affine space defined by the equation $f(x_1, \ldots, x_n) = 0$. Let $\mathbb{A}^n - Y$ be the set of points in $\mathbb{A}^n$ where $f \neq 0$; if $H \subseteq \mathbb{A}^{n+1}$ is the algebraic set defined by $H = \{(x_1, \ldots, x_{n+1}) \in \mathbb{A}^{n+1} : x_{n+1} f(x_1, \ldots, x_n) = 1\}$, then the rational map $(x_1, \ldots, x_n) \to (x_1, \ldots, x_n, f^{-1})$ sets the algebraic sets $\mathbb{A}^n - Y$ and H in birational correspondence. In particular $\mathbb{A}^n - Y$ is affine and its affine coordinate ring is the local ring $k[x_1, \ldots, x_n]_f$. If p is a point of X and Y is a hypersurface which does not contain p, then the above fact shows that $X - (X \cap Y)$ contains an open neighborhood U with $p \in U$ and U is birationally isomorphic to an affine open subset. This shows that any variety X has a base for the Zariski topology consisting of affine open subsets.

Let X and Y be any two varieties. Let $\varphi \colon X \to Y$ be a dominant rational map represented by the pair $\langle U, \varphi_V \rangle$. Let $f \in K(Y)$ be a rational function represented by the pair $\langle V, f \rangle$, where $V \subseteq Y$ is an open subset and f is a regular function on V. Since $\varphi_U(U)$ is a dense subset of Y, the subset $\varphi_U^{-1}(V)$ is non-empty and open in X. Therefore the composite $f \circ \varphi_U$ represents a regular function on $\varphi_U^{-1}(V)$, i.e. $f \circ \varphi$ is a rational function on X. This construction gives a bijection between the set of dominant rational maps from X to Y and the set of k-algebra homomorphisms from $K(Y)$ to $K(X)$. In particular two varieties X and Y are birationally isomorphic if and only if their function fields $K(X)$ and $K(Y)$ are isomorphic as k-algebras.

Definition 1.15 *A field extension K/k is separably generated if there is a transcendence base $\{x_i\}$ for K/k such that K is a separable algebraic extension of $k(\{x_i\})$. Such a transcendence base is called a separating transcendence base.*

A basic property of a finitely and separably generated field extension K/k is that any set of generators always contains a subset which is a separating transcendence base. A field K is called *perfect* if the polynomials in $k[x]$ are separable, i.e. the irreducible factors of any polynomial in $k[x]$ have distinct roots. For example if k is algebraically closed, then it is perfect. If

K/k is a finitely generated field extension over a perfect base field k, then it is separably generated. These results apply in particular to the function field K/k of an algebraic curve over the finite field of constants $k = \mathbb{F}_q$. The following result is the well-known theorem of the primitive element: if L is a finite separable extension of a field K, then there is an element α in L which generates L as an extension field of K. In fact if $\beta_1, \ldots, \beta_n$ is any set of generators of L over K, and if K is infinite, then α can be taken to be a linear combination $\alpha = c_1\beta_1 + \cdots + c_n\beta_n$ of the β_i with coefficients $c_i \in K$. An interesting application of this theorem is to the proof, which is now straightforward, that any variety of dimension r is birationally isomorphic to a hypersurface Y in $\mathbb{P}^{r+1}$.

1.1.6 Non-singular varieties

Let $Y \subseteq \mathbb{A}^n$ be an affine variety of dimension r, and let the ideal of Y be generated by the polynomials $f_1, \ldots, f_t \in k[x_1, \ldots, x_n]$, i.e. $I(Y) = (f_1, \ldots, f_t)$. Now the classical Jacobian criterion for a point $p \in Y$ to be simple is that the rank of the matrix $[(\partial f_j / \partial x_i)(p)]$ be $n - r$. Y is said to be non-singular if each of its points is simple. Recall that a noetherian local ring A with maximal ideal m and residue field $k = A/m$ always satisfies $\dim_k(m/m^2) \geq \dim A$; if strict equality holds, then A is called a *regular local ring*. This concept is of relevance in the study of non-singular varieties because if $Y \subseteq \mathbb{A}^n$ is an affine variety and $p \in Y$ is a point, then p is a simple point on Y if and only if the local ring $\mathcal{O}_{p,Y}$ is a regular local ring. This intrinsic characterization allows us to make the following definition.

Definition 1.16 *Let Y be any variety. Y is non-singular at a point $p \in Y$ if the local ring $\mathcal{O}_{p,Y}$ is a regular local ring. Y is non-singular if it is non-singular at every point.*

If Y is a variety then the set Sing Y of non-simple points of Y is a proper closed subset of Y. In particular if Y is a curve it can have at most a finite number of singular points. It is also clear that the projective line is non-singular. The completion of a local ring A with maximal ideal m can be defined as the projective limit ([2], p. 100)

$$\hat{A} = \varprojlim_n A/m^n.$$

If A is a noetherian local ring with maximal ideal m, then its completion $\hat{A}$ is a local ring with maximal ideal $\hat{m} = m \cdot \hat{A}$ and there is a natural injective homomorphism $A \to \hat{A}$. We also have $\dim A = \dim \hat{A}$. If M is a finitely generated A-module, its completion $\hat{M}$ with respect to the m-adic topology defined by the system of neighborhoods $\{m^n M\}_{n \in \mathbb{N}}$ is isomorphic to $M \otimes_A \hat{A}$. The completion $\hat{A}$ of a noetherian local ring A is regular if and only if A itself is regular. The completion of the local ring $k[x_1, \dots, x_n]$ at the maximal ideal $m = (x_1, \dots, x_n)$ is the ring $k[[x_1, \dots, x_n]]$ of formal power series over the residue field k. These rings play a distinguished role because of the Cohen Structure Theorem ([2], p. 123), which states that any complete regular local ring of dimension n containing some field is isomorphic to such a ring of formal power series.

Definition 1.17 *Two points $p \in X$ and $q \in Y$ are analytically isomorphic if there is an isomorphism $\hat{O}_p \cong \hat{O}_q$ as k-algebras.*

1.1.7 Smooth models of algebraic curves

In the dictionary between curves and function fields an important role is played by the concept of valuation which we now define

Definition 1.18 *Let K be a field and let G be a totally ordered abelian group. A valuation of K with values in G is a map $v: K - \{0\} \to G$ such that for all $x, y \in k$, $x, y \neq 0$, we have*

$$v(xy) = v(x) + v(y), \qquad v(x + y) \geq \min(v(x), v(y)).$$

The set $R = \{x \in K : v(x) \geq 0\} \cup \{0\}$ is called a *valuation ring* of v and the subset $m = \{x \in k : v(x) > 0\} \cup \{0\}$ is a maximal ideal of R, therefore R, m is a local ring. In general an integral domain in which R is the valuation ring of some valuation of its field of quotients is called a valuation ring. If R is a valuation ring and K is its field of quotients, we say that R is a valuation ring of K. If k is a subfield of K such that $v(x) = 0$ for all $x \in k - \{0\}$, then we say that v is a valuation of K/k, and R is a valuation ring of K/k.

In Section 1.2 we shall prove that if K is a field of algebraic functions of one variable over a field k and if R is a subring of K with $k \subset R \subset K$ and $m \subseteq R - \{1\}$ a non-zero ideal, then there exists a valuation ring R' of K with $R \subseteq R'$ and $m = M \cap R$, where M is the maximal ideal of R'. If A and B are local rings contained in a field K, we say that B dominates A if $A \subseteq B$ and $m_A = m_B \cap A$. It can be shown that a local ring R contained in a field K is a valuation ring if and only if it is a maximal element of the set of local rings contained in K with respect to the relation of domination. Also, the above fact can be used to show that every local ring contained in K is dominated by some valuation ring of K.

Definition 1.19 *A valuation v is discrete if its value group G is the set of integers. The corresponding valuation ring is called a discrete valuation ring, or d.v.r. for short.*

If A is a noetherian local domain of dimension one, with maximal ideal m, then the following are equivalent notions: (i) A is a discrete valuation ring, (ii) A is integrally closed, (iii) A is a regular local ring and (iv) m is a principal ideal. An integrally closed noetherian domain of dimension one is called a Dedekind domain; a typical example is $k[x]$. The localization of a Dedekind domain at a non-zero prime ideal is a discrete valuation ring. For example, if k is an arbitrary field and $f \in k[x]$ is an irreducible polynomial, then the localization $k[x]_f$ is a discrete valuation ring. In Section 1.2 we will show that all discrete valuation rings of $k(x)$ arise as the localizations $k[x]_f$ with f running over the set of all irreducible polynomials in $k[x]$ or $f = 1/x$. The integral closure of a Dedekind domain in a finite extension field of its quotient field is again a Dedekind domain.

In the following we assume that the field of constants k is algebraically closed. We have seen that if p is a point in a non-singular curve Y then its local ring $\mathcal{O}_p$ is a regular local ring of dimension one and hence it is a discrete valuation ring. Since $k \subseteq \mathcal{O}_p \subseteq K$, where K is the function field of Y, we see that $\mathcal{O}_p$ is actually a valuation ring of K/k.

Lemma 1.1 *Let Y be a quasi-projective variety, let $p, q \in Y$, and suppose that $\mathcal{O}_p = \mathcal{O}_q$ as subrings of $\mathcal{O}(Y)$. Then $p = q$.*

Proof. By embedding the quasi-projective variety Y in $\mathbb{P}^n$ for some n and replacing it by its closure, we may assume that Y is projective. After a suitable linear change of coordinates we may assume that the points p and q do not lie on the hyperplane H_0 defined by $x_0 = 0$. Since p and q lie on $Y \cap (\mathbb{P}^n - H_0)$ which is affine, way may assume without loss of generality that Y is an affine variety.

Now let A be the affine coordinate ring of Y. Then there are maximal ideals $m, n \subseteq \mathcal{A}$ such that $\mathcal{O}_p = A_m$ and $\mathcal{O}_q = A_n$. If $\mathcal{O}_q \subseteq \mathcal{O}_p$, then we must have $m \subseteq n$, but since m is maximal we have $m = n$ and therefore $p = q$.

The following result is a generalization of the simple fact that a rational function $k(x)$ has only a finite number of zeros and poles. A refinement of the result, which will be established in Section 1.2, will lead to the definition of principal divisor attached to a rational function.

Lemma 1.2 *Let K be a function field of dimension one over k, and let $x \in K$. Then x is contained in all valuation rings of K except possibly for a finite number.*

Proof. From the definition of valuation ring we verify that a necessary and sufficient condition for x to lie outside R is that $1/x \in m_R$. We may therefore rephrase the statement to read that if $y = 1/x$, then there are at most a finite number of valuation rings $R \subset K$ such that $y \in m_R$. Consider $k[y] \in K$ and observe that since k is algebraically closed, y is a transcendental element and hence $k[y]$ is a polynomial ring. Furthermore, since K is finitely generated and of transcendence degree 1 over k, K is a finite field extension of $k(y)$.

Let B be the integral closure of $k[y]$ in K and observe that B is a Dedekind domain and is a finitely generated k-algebra. If y is contained in a discrete valuation ring R of K/k, then $k[y] \subset R$ and since R is integrally closed in K, we have $B \subseteq R$. Let $n = m_R \cap B$. Then n is a maximal ideal of B and B is dominated by R. The integrally closed local ring B_n, obtained by localizing B at n, is a valuation ring of K/k and hence $B_n = R$ by the maximality of valuation rings.

Clearly if $y \in m_R$, then $y \in n$. Since B is integrally closed in K and finitely generated as a k-algebra, it is the coordinate ring of some affine variety Y. Since B is a Dedekind domain, Y is non-singular and of dimension 1. To say that $y \in n$ is the same as saying that y, as a regular function on Y, vanishes at the point of Y corresponding to n. But y is not identically zero, so it vanishes at only a finite set of points; these points are in one-to-one correspondence with the maximal ideals of B, and $R = B_n$ is determined by the maximal ideal n. Hence $y \in m_R$ for only finitely many valuation rings R of K/k.

The above argument actually gives a little bit more, since it shows that any discrete valuation ring of K/k is isomorphic to the local ring of a point on some non-singular curve.

The set of all discrete valuation rings of K/k will be denoted by C_K. In the following the elements of C_K will be called closed points; we will write $p \in C_K$, where p stands for the valuation ring R_p. Since C_K contains all the local rings of any non-singular curve with the function field K, and since k is assumed to be algebraically closed, C_K is infinite. We make C_K into a topological space by taking the closed sets to be the finite subsets and the whole space. If $U \subseteq C_K$ is an open subset of C_K, we define the ring of regular functions on U to be

$$\mathcal{O}(U) = \bigcap_{p \in U} R_p.$$

An element $f \in \mathcal{O}(U)$ defines a function from U to k by taking $f(p)$ to be the residue of f modulo the maximal ideal of R_p. If $f, g \in \mathcal{O}(U)$ take the same value at all points $p \in U$, then $f - g \in m_p$ for infinitely many p; this implies that $f = g$. Thus we can identify the elements of $\mathcal{O}(U)$ with functions from

U to k. By Lemma 1.2, any $f \in K$ is regular on some open set U. In particular the function field of C_K is K.

Definition 1.20 (i) *An abstract non-singular curve is an open subset $U \subseteq C_K$, where K is a function field of dimension 1 over k, with the induced topology and the induced notion of regular functions on its open subsets.*

(ii) *A morphism $\varphi: X \to Y$ between abstract non-singular curves or varieties is a continuous map such that for every open set $V \subseteq Y$, and every regular function $f: V \to k$, $f \circ \varphi$ is a regular function on $\varphi^{-1}(U)$.*

To give substance to the above definition let us first show that every non-singular quasi-projective curve Y is isomorphic to an abstract non-singular curve. Let K/k be the function field of Y and recall that the local ring $\mathcal{O}_p$ of each point $p \in Y$ is a discrete valuation ring. Furthermore if p and q are distinct points in Y, then $\mathcal{O}_p \neq \mathcal{O}_q$. Let $U \subseteq C_K$ be the set of all local rings of Y. We claim that U is an open subset of C_K; in fact we show that U contains a non-empty open set. Thus we may assume that Y is affine with affine coordinate ring A. In this case A is a finitely generated k-algebra and the function field K is the quotient field of A; the set U is the set of localizations of A at all its maximal ideals; all these local rings are discrete valuation rings; U consists of all discrete valuation rings of K/k containing A. Now let $x_1, \ldots, x_n$ be a set of generators of A over k. Then $A \subseteq R_p$ if and only if $x_1, \ldots, x_n \in R_p$. Thus $U = \bigcap_i U_i$, where $U_i = \{p \in C_K: x_i \in R_p\}$. But $\{p \in C_K: x_i \in R_p\}$ is a finite set. Therefore each U_i and hence also U is open. This shows that U is an abstract non-singular curve. It remains to show that the map $\rho: Y \to U$ given by $\varphi(p) = \mathcal{O}_p$ for each point $p \in Y$ is an isomorphism. For this we need only show that the regular functions on any open set are the same. This follows from the definition of the regular functions on U and the fact that for any open set $V \subseteq Y: \mathcal{O}(V) = \bigcap_{p \in V} \mathcal{O}_{p,Y}$. To establish the converse, we now prove the following important extension lemma.

Lemma 1.3 *Let X be an abstract non-singular curve, let $p \in X$, let Y be a projective variety, and let*

$$\varphi: X - \{p\} \to Y$$

be a morphism. Then there exists a unique morphism $\bar{\varphi}: X \to Y$ extending φ.

Proof. By embedding Y in some projective space $\mathbb{P}^n$, we see that it suffices to show that φ extends to a morphism of X into $\mathbb{P}^n$; if $\bar{\varphi}: X \to \mathbb{P}^n$ is the extension, then $\bar{\varphi}(X) \subset Y$.

Let $x_0, \ldots, x_n$ be homogeneous coordinates of $\mathbb{P}^n$ and define $U =$

$\{(x_0, \ldots, x_n) \in \mathbb{P}^n : x_0 x_1 \ldots x_n \neq 0\}$. By using induction on n we may assume that $\varphi(X - \{p\}) \cap U \neq \varnothing$. In fact, if $\varphi(X - \{p\})$ does not meet U, then it is contained in the union

$$\mathbb{P}^n - U = \bigcup_i H_i$$

of hyperplanes $H_i = \{(x_0, \ldots, x_n) : x_i \neq 0\}$, and since $\varphi(X - \{p\})$ is irreducible, it must be contained in some H_i, which in turn is isomorphic to $\mathbb{P}^{n-1}$; so the result would follow by induction. Hence in the following we assume that $\varphi(X - \{p\}) \cap U \neq \varnothing$.

The composition of φ with the map $(x_0, \ldots, x_n) \rightarrow x_i/x_j$ for each pair i, j defines a regular function f_{ij} on an open subset of X; in fact we obtain a regular function $f_{ij} \in K$ in the function field of X.

Let v be the valuation of K associated with the valuation ring R_p. Fix an integer k so that $v(f_{k0}) = \min_i v(f_{i0})$ and observe that

$$v(f_{ik}) = v(f_{i0}) - v(f_{k0}) \geq 0$$

for all i; hence $f_{0k}, \ldots, f_{nk} \in R_p$. Now define $\bar{\varphi}(p) = (f_{0k}(p), \ldots, f_{nk}(p))$ and $\bar{\varphi}(q) = \varphi(q)$ if $q \neq p$. We claim that $\bar{\varphi}$ is a morphism of X to $\mathbb{P}^n$ which extends φ, and that $\bar{\varphi}$ is unique. The uniqueness is clear by construction. To show that $\bar{\varphi}$ is a morphism, it will be sufficient to show that regular functions in a neighborhood of $\bar{\varphi}(p)$ pull back to regular functions on X. Let

$$U_k = \{(x_0, \ldots, x_n) \in \mathbb{P}^n : x_k \neq 0\};$$

since $f_{kk}(p) = 1$, $\bar{\varphi}(p) \in U_k$. Now U_k is affine with affine coordinate ring

$$k[x_0/x_k, \ldots, x_n/x_k].$$

These functions pull back to $f_{0k}, \ldots, f_{nk}$ which are regular at p by construction. It follows that for any smaller neighborhood V with $\bar{\varphi}(p) \in V \subseteq U_k$, regular functions on V pull back to regular functions on X. Hence $\bar{\varphi}$ is a morphism. This completes the proof.

The following important theorem is the main result of Section 1.1. It associates to a function field K of transcendence degree 1 over an algebraically closed field k a non-singular projective curve. This result establishes a dictionary between geometric concepts which are birationally attached to a curve and concepts pertaining to the function field of the curve. The result also establishes that every curve is birationally equivalent to a non-singular projective curve.

Theorem 1.1 (Existence of non-singular models.) *Let K be a function field of dimension 1 over k. Then the abstract non-singular curve C_K consisting of all valuation rings in K is isomorphic to a non-singular projective curve.*

Proof. Associated with the discrete valuation ring R_p of a point $p \in C_K$, there is a non-singular affine curve V and a point $q \in V$ with $R_p \cong \mathcal{O}_q$. The function field of V is K and V is isomorphic to an open subset of C_K. In particular every point $p \in C_K$ has an open neighborhood which is isomorphic to an affine variety. Since C_K is quasi-compact we can cover it with a finite number of open subsets U_i, each of which is isomorphic to an affine variety. Embed $U_i \subseteq \mathbb{A}^{n(i)}$ and think of $\mathbb{A}^{n(i)}$ as an open subset of $\mathbb{P}^{n(i)}$ and let Y_i be the closure of U_i in $\mathbb{P}^{n(i)}$. Then Y_i is a projective variety, and we have a morphism $\varphi_i: U_i \to Y_i$, which is an isomorphism of U_i onto its image. The complement of U_i in C_k is a finite set and Lemma 1.3 guarantees the existence of a morphism $\bar{\varphi}_i: C_K \to Y_i$ extending φ_i. Via the Segre embedding (Shafarevitch [76]) we see that the product of the projective varieties Y_i is a projective variety $\prod_i Y_i$. Let $\varphi: C_K \to \prod_i Y_i$ be the diagonal map $\varphi(p) = \prod_i \bar{\varphi}_i(p)$ and let Y be the closure of the image of φ. Y is a projective variety and $\varphi(C_K)$ is dense in Y. Therefore Y is a curve. To show that φ is an isomorphism, we recall that the U_i cover C_K and that each point $p \in C_K$ belongs to some U_i. Also there is a commutative diagram

of dominant morphisms, where π is the projection map onto the i-th factor. If p is a point in C_K we have natural inclusions of local rings

$$\mathcal{O}_{\varphi_i(p), Y_i} \hookrightarrow \mathcal{O}_{\varphi(p), Y} \hookrightarrow \mathcal{O}_{p, C_k}.$$

Now, the local rings at both ends are isomorphic and hence are isomorphic to the middle one. Thus for any $p \in C_K$ the map $\varphi_p^*: \mathcal{O}_{\varphi(p), Y} \to \mathcal{O}_{p, C_k}$ is an isomorphism. Next let q be any point of Y. Then $\mathcal{O}_q$ is dominated by some discrete valuation ring R of K/k, e.g. a localization of the integral closure of $\mathcal{O}_q$ at a maximal ideal. On the other hand $R = R_p$ for some $p \in C_K$, and $\mathcal{O}_{\varphi(p)} \cong R$, so by Lemma 1.1 we must have $q = \varphi(p)$. This shows φ is surjective. Also φ is clearly injective, because distinct points of C_K correspond to distinct subrings of K. Thus φ is a bijective morphism of C_K to Y, and for every $p \in C_K$ the induced map φ_p^* is an isomorphism, so φ itself is an isomorphism. This completes the proof of the theorem.

1.2 Algebraic aspects

1.2.1 Introduction

Let k be a field. By an algebraic function field of one variable over k we mean a field K which contains k as a subfield and satisfies the condition:

K contains an element x which is transcendental over k and K is a finite algebraic extension of $k(x)$. For some of the applications k may be taken to be a perfect field; in particular k can be the finite field $\mathbb{F}_q$ of q elements or the separable closure of $\mathbb{F}_q$. The set of elements in K which are algebraic over k form a subfield k' of K which is called the *field of constants* of K. In the following we make the assumption that k is already the field of constants of K, i.e. k is algebraically closed in K.

1.2.2 Points on the projective line $\mathbb{P}^1$

Let $K = k(x)$ be a pure transcendental extension of the field k, i.e. x is transcendental over k. A basic step in the determination of all the points on an algebraic curve C defined over k is to realize C as a covering of the projective line $\mathbb{P}^1$; one then determines all the points of $\mathbb{P}^1$, these latter being thought of as valuation rings in the rational function field $k(x)$; finally, by purely algebraic methods, it is shown that over each point of $\mathbb{P}^1$ there lies a finite number of points in C, i.e. a valuation ring of $k(x)$ has a finite number of extensions to valuation rings of the function field of C.

To determine all valuation rings of $K = k(x)$ one proceeds as follows. Let f be an irreducible polynomial in $k[x]$. Now any element $u \in K$ can be written in the form $u = g/h$ with $g, h \in k[x]$. Consider the subring of K defined by

$$A_f = \{g/h \in K : h \text{ is not divisible by } f\}.$$

Clearly $k \subseteq A_f \subseteq k[x]$, and $A_f \neq k[x]$ because $1/f \notin A_f$. The set $m_f = fA_f$ is a maximal ideal of A_f. Furthermore if u is an element in K with $u \notin A_f$, then $1/u \in A_f$. This shows then that the local ring (A_f, m_f) is in fact a valuation ring. We have thus associated to every irreducible polynomial a valuation ring (A_f, m_f). If f and f' are distinct irreducible polynomials, then the valuation rings A_f and $A_{f'}$ are distinct because f^{-1} belongs to $A_{f'}$ but not to A_f.

If we set $y = 1/x$, then $k(y) = k(x)$. To every irreducible polynomial in y there is associated a valuation ring of $k(x)$. This applies in particular to $y = x^{-1}$; we denote by $(A_{1/x}, m_{1/x})$ the corresponding valuation ring. The ring $A_{1/x}$ is actually distinct from all the rings A_f defined above because if f is any irreducible polynomial in x, we have $x \in A_f$ while $x \notin A_{1/x}$.

The valuations associated to the above valuation rings are defined as follows. If $u \in k(x)$ has a representation of the form $u = f^v \cdot g/h$, where g and h are not divisible by f, then the valuation corresponding to f is given by the order function

$$\mathrm{ord}_f(u) = v.$$

To obtain the valuation associated to $A_{1/x}$ we write an element $u \in k(x)$ in the form

$$u = \frac{a_0 + a_1 x + \cdots + a_m x^m}{b_0 + b_1 x + \cdots + b_n x^n}, \qquad (a_m b_n \neq 0)$$

$$= \left(\frac{1}{x}\right)^{n-m} u',$$

where u' is the quotient of two polynomials in $1/x$ both of which are not divisible by $1/x$. In this case we put

$$\text{ord}_{1/x}(u) = n - m.$$

We assert that the valuation rings A_f (for all irreducible polynomials $f \in k[x]$) and $A_{1/x}$ exhaust all the valuation rings of the rational function field $k(x)$. Let A be any valuation ring in $k(x)$ and let m be its maximal ideal. If $x \in A$, then since $k \subset A$, we have $k[x] \subset A$. Since m is a prime ideal in A, $m \cap k[x]$ is also a prime ideal in $k[x]$ and is therefore equal to the zero ideal or to a principal ideal in $k[x]$ generated by some irreducible polynomial f. The first case is impossible because every non-zero element of $k(x)$ would belong to $A - m$, i.e. it would be a unit, and this would imply that $A = k(x)$. Hence only the second case is possible, i.e. $m \cap k[x] = (f)$. If g and $h \in k[x]$ and h is not divisible by f, then h^{-1} is a unit in A_f and $g/h \in A$. This implies that $A_f \subseteq A$. Let u be an element of $k(x)$ not contained in A_f; then we may write $u = g/h$ where g and h have no factor in common and h is a multiple of f. If u were in A, then $h^{-1} = g^{-1} \cdot u$ would also belong to A; but this is impossible because $h \in m_f$ and hence h is not a unit in A. This shows that if $x \in A$, then $A = A_f$ for some irreducible polynomial f. If x is not in A, then $x' = x^{-1}$ is and $m \cap k[x']$ is a principal ideal generated by an irreducible polynomial $f'(x') \in k[x']$. Since x' is not a unit in A, it is in $m \cap k[x']$ and is therefore divisible by $f'(x')$ in $k[x']$. This can only happen if we have $f' = ax'$, $a \in k^x$. Thus we may assume $f' = x'$ and that $A = A_{1/x}$.

Remark It is well known that the number of irreducible polynomials in $\mathbb{F}_q[x]$ of degree m is

$$N_m = m^{-1} \sum_{d \mid m} \mu(m/d) q^d,$$

where μ is the Mobius function.

1.2.3 Extensions of valuation rings

In order to view an algebraic curve C as a covering of the projective line $\mathbb{P}^1$ we must study the problem of how to extend valuation rings in the function field of $\mathbb{P}^1$ to the function field of the curve C. This is the aim of the following theorem.

Theorem 1.2 *Let K be the function field of a curve C; let k be the field of constants. Let A be a subring of K containing k and m_A an ideal in A not containing 1 and different from the zero ideal. Then there exists a valuation ring B of K with maximal ideal m_B and satisfying $m_A \subset m_B \cap A$.*

Proof. If A' is any subring of K containing A, we put

$$m_A \cdot A' = \{m \cdot a : m \in m_A, a \in A'\};$$

$m_A \cdot A'$ is an ideal. Consider the family of subrings of K

$$\mathcal{F} = \{A' \text{ subring of } K : A \subset A' \text{ and } m_A \cdot A' \neq A'\}.$$

$\mathcal{F}$ is not empty since A itself belongs to it. We want to apply Zorn's lemma to show that $\mathcal{F}$ has at least one maximal element. We will then show that such a maximal element is a valuation ring.

To prove that $\mathcal{F}$ is an inductive family we need verify that if $\mathcal{F}'$ is a non-empty subfamily of $\mathcal{F}$ with the property that of any two members of $\mathcal{F}'$ one contains the other, then $\mathcal{F}$ contains a ring which contains all the rings of the family $\mathcal{F}'$. Let us denote by A_1 the set theoretic union of all the rings in $\mathcal{F}'$. If x and $y \in A_1$, then $x \in A'$, $y \in A''$, where A' and A'' are in $\mathcal{F}'$ and one of the rings A', A'' contains the other. For instance if A' contains A'', then x and y are both in A', and hence $x - y \in A'$, $xy \in A'$, which shows that $x - y$ and xy are in A_1. Clearly the same is true if instead we have $A' \subset A''$; therefore A_1 is a ring. Since every ring belonging to $\mathcal{F}$ contains A, A_1 contains A. We assert that $m_A \cdot A_1 \neq A_1$. In fact if this were not the case then we could represent 1 in the form $1 = x_1 y_1 + \cdots + x_h y_h$, with $x_i \in m_A$, $y_i \in A_1$ $(1 \leq i \leq h)$. Each y_i would belong to some ring $A^{(i)} \in \mathcal{F}'$. For each pair (i, j) one of the rings $A^{(i)}$, $A^{(j)}$ would contain the other; since there are only a finite number of rings $A^{(i)}$, it follows that they would all be contained in one of them, say $A^{(k)}$. But then we would have $1 = \sum_{i=1}^{h} x_i y_i \in m_A \cdot A^{(k)}$, therefore $m_A \cdot A^{(k)} = A^{(k)}$, which is impossible since $A^{(k)} \in \mathcal{F}'$. Thus we have $m_A \cdot A_1 \neq A_1$; this implies $A_1 \in \mathcal{F}$, and hence $\mathcal{F}$ is inductive.

Let B be a maximal ring in $\mathcal{F}$. We claim that B is a valuation ring of K. Consider the set

$$S = \{\beta \in B : \beta - 1 \in m_A \cdot B\};$$

it is clearly closed under multiplication, and the localization of B at S is the ring

$$B_S = \{\beta s^{-1} : \beta \in B, s \in S\}.$$

We have $B \subseteq B_S$. We show first that $1 \notin m_A \cdot B_S$. Suppose for a moment that we do have

$$1 = \sum_{i=1}^{h} \alpha_i \beta_i s_i^{-1}, \qquad \alpha_i \in m_A, \; \beta_i \in B, \qquad s_i \in S, \qquad (1 \le i \le h).$$

Set $s = s_1 \cdots s_h$; then $s \in S$ and $s = 1 + \sum_{i=1}^{h} \alpha_i' \beta_i'$, with $\alpha_i' \in m_A$, $\beta_i' \in B$, and we have

$$1 = \sum_{i=1}^{h} \alpha_i \left(\prod_{j \ne i} s_j \right) \beta_i - \sum_{i=1}^{h} \alpha_i' \beta_i' \in m_A \cdot B,$$

which is impossible. Thus B_S belongs to $\mathscr{F}$ and contains B; hence since B is maximal we must have $B_S = B$; this implies that if $s \in S$, then $s^{-1} \in B$. Now let u be any element of K not in B; then $B \ne B[u]$ and therefore $B[u] \notin \mathscr{F}$ and $m_A B[u] = B[u]$. Hence 1 may be represented in the form $1 = \sum_{i=1}^{n} \alpha_i u^i$, with $\alpha_i \in m_A \cdot B$, $(1 \le i \le n)$. Since $1 - \alpha_0 \in S$, we also write $1 = \sum_{i=1}^{n} \alpha_i' \cdot u^i$ with $\alpha_i' = \alpha_i (1 - \alpha_0)^{-1} \in m_A \cdot B$. We may furthermore assume that, among all representations of 1 in this form, we have selected the one with the minimum number of terms, i.e. it is impossible to represent 1 in the form $1 = \sum_{i=1}^{n'} \alpha_i'' u^i$ with $n' < n$, $\alpha_i'' \in m_A \cdot B$ for $i \le i \le n'$. Let us assume for a moment that $u^{-1} \notin B$. Then exactly as before 1 can be represented in the form $1 = \sum_{i=1}^{m} \beta_i' u^{-i}$ with $\beta_i' \in m_A \cdot B$, $(1 \le i \le m)$; furthermore, we may assume that this is a representation with the minimum number of summands. If $n \ge m$, we may write $u^n = \sum_{i=1}^{m} \beta_i' u^{n-i}$, therefore

$$1 = \sum_{i=1}^{n-1} \alpha_i' u^i + \alpha_n' \left(\sum_{i=1}^{m} \beta_i' u^{n-i} \right),$$

which is impossible by virtue of our choice of n. Exchanging the roles played by u and u^{-1}, we see in the same way that the assumption that $n \le m$ also leads to a contradiction. This shows that $u^{-1} \notin B$ is impossible and hence obviously $u^{-1} \in B$. Now, since $m_A \cdot B \ne B$, we have $B \ne K$; therefore B is a valuation ring.

Let m_B be the ideal of non-units of B; m_B is a maximal ideal of B. Since $m_A \cdot B \ne B$, an element of m_A cannot be a unit in B, and therefore $m_A \subset m_B \cap A$. This completes the proof of Theorem 1.2.

1.2.4 Points on a smooth curve

If x is a non-constant element of a field K/k of algebraic functions of one variable, then we can think of $k(x)$ as the function field of the projective line $\mathbb{P}^1$. If A is a valuation ring in K with $k \subset A$, then the intersection $A \cap k(x)$ is a valuation ring of $k(x)$ and hence corresponds to a point of $\mathbb{P}^1$.

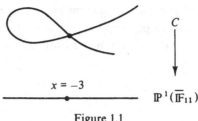

Figure 1.1

Conversely, to a point in $\mathbb{P}^1$ there corresponds a valuation ring in $k(x)$ and as we have seen in Section 1.2.3 such a valuation ring can be extended to a valuation ring A of K. With A there is associated a smooth point on a curve C with function field K. If K is of degree n over $k(x)$, then we can think of it as the function field of a curve C which is a covering of the projective line $\mathbb{P}^1$ of degree n. Heuristically this gives us an idea of where the points of C come from. Since $\mathbb{P}^1$ has an infinite number of points, so does the curve C.

Example Let C be the curve $y^2 - y = x^3 - x^2$ and let $K = \bar{\mathbb{F}}_{11}(x, y)$ be its function field. As indicated in Figure 1.1, above each point on $\mathbb{P}^1$ there are two points on C except over $x = -3$ where ramification occurs. In order to attach to each point p of C_K an order function, we prove the following result. Let A be a valuation ring of K and m its maximal ideal.

Lemma 1.4 The ring A contains an element t such that $m = t \cdot A$ and $\bigcap_n t^n \cdot A = \{0\}$.

Proof. Let x be a non-zero element of m. Assume there is a finite set of elements $\{t_0, \ldots, t_e\}$ which satisfy the condition:

(i) $t_0 = x, \; t_i/t_{i+1} \in m$ for $0 \le i \le e - 1, \quad t_e = 1.$

Let $\{u_1, \ldots, u_d\}$ be elements in A whose residue classes $\bar{u}_1, \ldots, \bar{u}_d$ modulo m are linearly independent over k. We want to show that the ed elements $t_i u_j, \; 1 \le i \le e, \; 1 \le j \le d$, of K are linearly independent over $k(x)$. Suppose on the contrary that there is a relation of the form

(ii) $\sum_{i=0}^{e} \sum_{j=0}^{d} a_{ij}(x) t_i u_j = 0,$

where the coefficients $a_{ij}(x) \in k(x)$ are not all zero. After multiplying by the common denominator of the rational functions $a_{ij}(x)$ and dividing through by a suitable power of x if necessary, we may suppose that $a_{ij}(x)$ are polynomials and that at least one of them has a non-zero constant term $a_{ij} = a_{ij}(0)$; observe that $a_{ij}(x) - a_{ij} = xb_{ij}(x) \in X \cdot A$, where $b_{ij}(x) \in k[x]$. Let h be the index ≥ 1 of the last non-zero row of the e by d matrix (a_{ij}). The equation (ii) gives

$$\sum_{i=1}^{h} \sum_{j=1}^{d} a_{ij} t_i u_j = xw,$$

with $w = -\sum_{i=1}^{e} \sum_{j=1}^{d} b_{ij}(x) t_i u_j \in A$. Rewrite the above sum in the form

$$\sum_{j=1}^{d} a_{hj} u_j = wx/t_h - \sum_{i=1}^{h-1} \sum_{j=1}^{d} a_{ij}(t_i/t_h) u_j.$$

Since $h \geq 1$, we have $x/t_h \in m$; if $i < h$, then t_i/t_h also belongs to m. Hence the right-hand side of the above expression belongs to m and $\sum_{j=1}^{d} a_{hj} \bar{u}_j = 0$. But this is impossible, since at least one of the $a_{kj} \neq 0$ and $\bar{u}_1, \ldots, \bar{u}_d$ are linearly independent over k. Therefore the elements $\{t_i u_j : 1 \leq i \leq e, 1 \leq j \leq d\}$ are linearly independent over $k(x)$ as asserted.

Since $x \in m$, x is not a constant; the field K is therefore a finite algebraic extension of $k(x)$ and if $n = [K : k(x)]$, then $ed \leq n$. Taking $e = 1$, $t_0 = x$ and $t_1 = 1$ we get a set that satisfies the condition (i) and therefore we can conclude that $d \leq n$. This implies that the residue field A/m cannot contain more than n linearly independent elements over k. This proves that A/m is an algebraic extension of finite degree over k.

Next, taking $d = 1, u_1 = 1$, we see that the number of terms in a sequence $\{t_0, \ldots, t_e\}$ which satisfies the condition (i) above is bounded by $n + 1$. Now from the family of all sequences that satisfy the condition (i), take one, say $\{t_0, t_1, \ldots, t_e\}$, with the largest possible number of terms, and set $t = t_e$. Let z be any element of m; we claim that $z \in tA$. If this were not the case, z/t would not be an element of A and therefore t/z would be a non-unit of A, i.e. $t/z \in m$. But this would imply that the sequence $\{t_0, \ldots, t_{e-1}, t_{e-1}/z, t_e\}$ containing $e + 2$ elements also satisfies condition (i), which is impossible. This proves that $m = t \cdot A$ is a principal ideal. Let y be a non-zero element of m; if $y \in t^m \cdot A$, set $u_1 = y, u_{i+1} = y/t^i, 1 \leq i \leq m - 1, u_{m+1} = 1$. Then we have $u_i/u_{i+1} \in m, (1 \leq i \leq m)$; from what we have already shown it follows that $m \leq [K : k(y)]$. This implies that $\bigcap_{n=0}^{\infty} t^n \cdot A = \{0\}$ and the proof of the lemma is complete.

As indicated above, the argument also shows that the residue class field of the valuation ring A, m is a field of finite degree over the field of constants k. This leads to the following important definition.

Definition 1.21 *Let K be a field of algebraic functions in one variable with k as the field of constants. Let p be a point of the smooth nonsingular model*

C_K and let A, m be the corresponding valuation ring and maximal ideal. The degree of p is defined to be

$$\deg(p) = [k_p : k],$$

where $k_p = A/m$ is the residue class field. To simplify notation we will often write $d(p)$ instead of $\deg(p)$.

Lemma 1.4 shows that a valuation ring in a field of algebraic functions of one variable is a regular local ring and hence is the local ring of a smooth point on some algebraic curve. The element t which generates the maximal ideal of the valuation ring corresponding to the point p is called a *local uniformizing parameter* at p.

Let A be the valuation ring of a point p and m its maximal ideal. To define an order function, i.e. an exponential valuation, associated to A, we put for each non-zero element $f \in K$,

$$v_p(f) = n,$$

where n is the largest integer such that $u \in t^n A$, where t is a local uniformizing parameter at p. To show that $v_p(f)$ is well defined we observe that if t_0 is another local uniformizing parameter at p, then we have $m = t_0 \cdot A$ and $t \in t_0 \cdot A$; hence $t = t_0 \cdot u$ with some $u \in A$. Since $t_0 \in tA$, $u^{-1} \in A$, it follows that $t^n \cdot A \subset t_0^n \cdot A$ and $t_0^n \cdot A \subset t^n \cdot A$ for every n. This proves that $v_p(f)$ is independent of the element t used to define it.

If u and w are non-zero elements in K, then

$$v_p(uw) = v_p(u) + v_p(w),$$

and if $u + w \neq 0$, then

$$v_p(u + w) \geq \min\{v_p(u), v_p(w)\}.$$

This last inequality can be replaced by an equality if $v_p(u) \neq v_p(w)$. More generally, we see by induction on h that if $x_1, \ldots, x_h$ are any nonzero elements in K, with $x_1 + \cdots + x_h \neq 0$, then

$$v_p(x_1 + \cdots + x_h) \geq \min_i v_p(x_i),$$

with equality holding if there is only one index j such that $v_p(x_j) = \min_i\{v_p(x_i)\}$. Often for convenience we write $v_p(0) = \infty$.

1.2.5 Independence of valuations

The following theorem will play a key role in the proof of the Riemann–Roch theorem in Chapter 2. It allows one to ascertain the existence of a

rational function on a smooth curve with zeros and poles at a finite set of
points with preassigned orders.

Theorem 1.3 *Let K be the function field of a curve, and let $p_1, \ldots, p_h$ be h
distinct points of C_K. Assign to each p_i an element u_i of K and some integer
m_i. Then there exists an element u of K which satisfies the h conditions
$v_i(u - u_i) \geq m_i$ with $v_i = v_{p_i}$.*

Proof. We use induction on the number of points. The case $h = 1$ is clear.
Assume $h > 1$ and that the theorem holds for systems of $h - 1$ points. Then,
given $h - 1$ integers $e_1, \ldots, e_{h-1}$, there always exists an element u of K such
that $v_i(u) = e_i$, $1 \leq i \leq h - 1$. To see this we note that there exist elements
w_i with $v_i(w_i) = e_i$, $1 \leq i \leq h - 1$, and by the induction hypothesis there
exists an element u such that $v_i(u - w_i) \geq e_i + 1$. Since $u = (u - w_i) + w_i$,
we then have $v_i(u) = \min\{v_i(u - w_i), v_i(w_i)\} = e_i$.

   We claim that the valuations $v_1, \ldots, v_h$ corresponding to the points
$p_1, \ldots, p_h$ are independent, i.e. no linear relation of the form

(iii) $$v_h(z) = \sum_{i=1}^{h-1} r_i v_i(z), \qquad r_i \in \mathbb{Q},$$

holds for every non-zero element $z \in K$. Suppose on the contrary that such
a relation does exist, and consider first the case where at least one of the
rational numbers r_i is < 0. The induction hypothesis and the previous
argument show that there exist elements z and z' in K such that

$$v_i(z) = 1, \qquad v_i(z') = 0 \qquad \text{if} \qquad r_i \geq 0,$$

$$v_i(z) = 0, \qquad v_i(z') = 1 \qquad \text{if} \qquad r_i < 0.$$

From (iii) we see that $v_h(z) \geq 0$, $v_h(z') < 0$. Now, since $v_i(z) \neq v_i(z')$ we have
$v_i(z + z') = \min\{v_i(z), v_i(z')\} = 0, (1 \leq i \leq h - 1)$; this implies $v_h(z + z') = 0$;
but this is impossible because $v_h(z') < v_h(z)$ and hence $v_h(z + z') = v_h(z') < 0$.
We now consider the case where all the rational numbers $r_i \geq 0$. Since not
all r_i are zero we may assume $r_1 > 0$. Let us observe that not all $r_2, \ldots,$
$r_{h-1} = 0$, for if this were the case, then the relation (iii) implies the two
valuation rings

$$A_1 = \{z \in K : v_1(z) \geq 0\} \qquad \text{and} \qquad A_h = \{z \in K : v_h(z) \geq 0\}$$

are identical and the points $p_1 = p_h$ which is impossible. Since at least one
of the numbers $r_2, \ldots, r_{h-1}$ is > 0 we can replace the linear relation (iii) by

$$v_1(z) = r_1^{-1} v_h(z) - \sum_{i=2}^{h-1} r_1^{-1} r_i v_i(z),$$

and apply to it the earlier argument. This establishes the independence of $v_1, \ldots, v_h$.

We now use the fact that there does not exist a system of h rational numbers $r_1, \ldots, r_h$ not all zero such that $\sum_{i=1}^{h} r_i v_i(z) = 0$ for all $z \in K^\times$ in order to show the existence of h elements $z_1, \ldots, z_h \in K^\times$ such that $\det(v_i(z_j)) \neq 0$. To show this we shall construct these elements step by step. For z_1 we take any element in $K^\times$ with $v_1(z_1) \neq 0$. We have

$$\dim_Q \left\{ (r_1, \ldots, r_h) \in Q^h : \sum_{i=1}^{h} r_i v_i(z_1) = 0 \right\} = h - 1.$$

Assume we have already determined $z_1, \ldots, z_k$, $(k < h)$, in such a way that

$$\dim_Q \left\{ (r_1, \ldots, r_h) \in Q^h : \sum_{i=1}^{h} r_i v_i(z_j) = 0, 1 \leq j \leq k \right\} \leq h - k;$$

let $(r_1, \ldots, r_h)$ be a non-zero vector in the above vector space and determine z_{k+1} so that $\sum_{i=1}^{h} r_i v_i(z_{k+1}) \neq 0$. Now the vector space

$$\left\{ (r_1, \ldots, r_h) \in Q^h : \sum_{i=1}^{h} r_i v_i(z_j) = 0, 1 \leq j \leq k + 1 \right\}$$

has Q-dimension $\leq h - (k + 1)$. Continuing in this way we obtain elements $z_1, \ldots, z_h$ in $K^\times$ such that no non-trivial linear combination of the valuations $v_1, \ldots, v_h$ with rational coefficients vanishes at $z_1, \ldots, z_h$. Since the values $v_i(z_j)$ are rational we obtain $\det(v_i(z_i)) \neq 0$.

The system of equations

$$\sum_{j=1}^{h} c_{j,k} v_i(z_j) = \begin{cases} -1 & \text{for} \quad i = k \\ 1 & \text{for} \quad i \neq k, \end{cases}$$

where k is any index from 1 to h, has a solution $(c_{1k}, \ldots, c_{hk})$ in rational numbers. Let d be an integer such that the numbers $dc_{ik}, 1 \leq i \leq h, 1 \leq k \leq h$ are all integral. If we set $y_k = \prod_{j=1}^{h} z_j^{dc_{jk}}$, we have

$$v_i(y_k) = \begin{cases} -d & \text{for} \quad i = k \\ d & \text{for} \quad i \neq k. \end{cases}$$

Put $x_k = (1 + y_k^{-1})^{-1}$. If $i \neq k$, we have $v_i(y_k^{-1}) < 0 = v_i(1)$, therefore $v_i(1 + y_k^{-1}) = v_i(y_k^{-1}) = -d$, and $v_i(x_k) = d$. On the other hand, $x_k - 1 = -y_k^{-1}(1 - y_k^{-1})^{-1}$ and $v_k(y_k^{-1}) = d$, therefore $v_k(x_k - 1) = d$. Since we are free to choose d as large as desired, we select it so that

$$d + v_i(w_j) \geq \max\{m_1, \ldots, m_h\}, \qquad 1 \leq i, j \leq h,$$

and we set

$$u = \sum_{i=1}^{h} x_i w_i.$$

Then $u - w_i = (x_i - 1)w_i + \sum_{k \neq i} x_k w_k$, and we have $v_i((x_i - 1)w_i) = d + v_i(w_i) \geq m_i$, and if $k \neq i$, $v_i(x_k w_k) = d + v_i(w_k) \geq m_i$ from which it follows that $v_i(u - w_i) \geq m_i$ ($1 \leq i \leq h$). The theorem is thus proved if we take $u_i = w_i$.

The above proof also also establishes the following corollary.

Corollary 1.3.1 *Let S be a finite set of distinct points on a smooth curve C_K with function field K. Assign to each point $P \in S$ an integer m_P. Then there is a rational function $f \in K$ which satisfies the conditions*

$$\mathrm{ord}_P(f) = m_P$$

for all points $P \in S$.

Exercises

1. Verify that the algebraic curve $xy^3 + yz^3 + zx^3$ is nonsingular over any field of characteristic different from 7. Do the same for the curve $y^2 - y = x^3 - x^2$ and the field $\mathbb{F}_{11}$ and determine the nature of the singularity. Verify that the curve $y^2 - y = x^3$ is non-singular over a field of characteristic 2.

2. (Automorphism groups of algebraic curves) It is well known that over the field of complex numbers the full automorphism group of the Klein curve $xy^3 + yz^3 + zx^3$ is $PSL_2(\mathbb{F}_7)$. Verify that the same is true over the algebraic closure of any finite field $\mathbb{F}_p$, $p \neq 7$. What can you say about the automorphism group of the singular curve over the finite field $\mathbb{F}_7$?

3. Show that the automorphism group of the curve $y^2 - y = x^3$ is isomorphic to $SL_2(\mathbb{F}_3)$. (*Hint:* Use the fact that $SL_2(\mathbb{F}_3)$ is isomorphic to the unit group of the ring of integral quaternions of Hurwitz.)

4. Consider the plane projective curve $x^4 + y^4 + z^4$ over a field $\mathbb{F}$ of characteristic different from 2. Let $G_1 = S_3$ be the group of permutations of the coordinates in P^2 and G_2 the group generated by multiplication of coordinates by roots of unity. Show that the automorphism group of the given curve is the subgroup of $PGL_2(\mathbb{F})$ of order 96 generated by G_1 and G_2.

5. Let $SL_2(\mathbb{F}_p)$ act by a linear change of coordinates on the field $\mathbb{F}_p(x, y)$. Define two elements by

$$L = x^p y - xy^p, \qquad Q = \frac{x^{p^2}y - xy^{p^2}}{L}.$$

Prove that any rational integral invariant I, with integral coefficients, of the group $SL_2(\mathbb{F}_p)$ is a rational integral function of L and Q with integral coefficients. (Dickson, *On Invariants and the Theory of Numbers*, Dover, New York, page 38.)

6. Let k be the algebraic closure of the finite field $\mathbb{F}_q$ and define a group $H = \{\lambda \in k^x: \lambda^{q+1} = 1\}$. Drienfeld has studied the action of $SL_2(\mathbb{F}_q)$ by linear change of coordinates and of H by homothety on the affine irreducible curve

$$\mathscr{C} = \{(x, y) \in k^2: xy^q - x^q y = 1\}.$$

Show how to extend this action to the non-singular projective curve

$$\bar{\mathscr{C}} = \{(x, y, z): xy^q - x^q y = z^{q+1}\}.$$

Show that over the field $\mathbb{F}_{q^2}$ the curve $\bar{\mathscr{C}}$ can be written as $x^{q+1} + y^{q+1} + z^{q+1}$ on which the unitary group $U_3(\mathbb{F}_q)$ acts naturally. What relation is there between these two group actions?

7. The general theory of abelian varieties with complex multiplications applied to the Jacobian of the Klein curve $xy^3 + yz^3 + zx^3$ implies that there is a simple formula for counting the number of points on the curve which are rational over the finite field $\mathbb{F}_p$ in terms of how the rational prime p decomposes in the complex quadratic field of discriminant -7. Write explicitly such a formula. (*Hint*: Use the method of Delsarte, Nombre de solutions des equations polynomials sur un corps fini, *Seminaire Bourbaki* No. 39 (1950).)

Notes

There are many excellent introductions to the theory of algebraic curves from the point of view of the algebraic geometer: Hartshorne's *Algebraic Geometry*, Shafarevitch's *Basic Algebraic Geometry*, Fulton's *Algebraic Curves*. From the point of view of number theory there is no suitable introduction and the presentation in this first chapter is only a compendium of the most basic definitions and results taken from the standard literature on algebraic geometry and the classic arithmetic treatment in C. Chevalley, *Introduction to the Theory of Functions of One Variable*. Excellent survey articles have been written by J. Tate (Arithmetic of elliptic curves, *Inventiones Math.*, vol. 23 (1974)) and B. Mazur (Arithmetic on Curves, Bull. A.M.S., vol. 14 (1986)). For a brief survey of the classical literature it may be worthwhile to consult J. Dieudonné's *History of Algebraic Geometry* (Wadsworth Adv. Books and Software (1985)).

2

The Riemann–Roch theorem

2.1 Divisors

Throughout this chapter K will denote a field of algebraic functions of one variable with k as the exact field of constants. For applications to number theory and coding theory it will be useful to specialize k as the finite field $\mathbb{F}_q$ or its algebraic closure $\overline{\mathbb{F}}_q = \bigcup_{n=1}^{\infty} \mathbb{F}_{q^n}$. For notational as well as linguistic convenience we shall use the curve $C = C_K$ with function field K whose existence is established by Theorem 1.1 in Chapter 1; with this dictionary in mind, we shall refer interchangeably to the discrete valuation rings of K as the closed points of C.

Definition 2.1 *A divisor of K is an element of the free Abelian group generated by the set of closed points of C.*

The group of divisors of K is denoted by $\mathrm{Div}(C)$; the group operation will be written additively. Thus any element D in $\mathrm{Div}(C)$ has the form

$$D = \sum_P \mathrm{ord}_P(D)P,$$

where the sum is taken over all closed points P of C, and the coefficients $\mathrm{ord}_P(D)$ are integers all of which are zero except for a finite number. The support of D, denoted by $\mathrm{supp}(D)$, is the set of P with $\mathrm{ord}_P(D) \neq 0$. A divisor D is called integral, or more suggestively positive, if $\mathrm{ord}_P(A) \geq 0$ for all closed points P; a divisor D_1 is said to divide a divisor D_2 if $D_2 - D_1$ is positive; when this is the case we write $D_2 \geq D_1$. Two divisors D_1 and D_2 are said to be relatively prime if $\mathrm{ord}_P(D_1) \neq 0$ implies that $\mathrm{ord}_P(D_2) = 0$. The *degree* $d(D)$ of a divisor D is defined by the formula

$$d(D) = \sum_P \mathrm{ord}_P(D) \cdot d_P,$$

where the sum is taken over all closed points of K; the integer d_P is the degree of the residue field $k_P = R_P/m_P$ over the field of constants k; here

R_P is the valuation ring associated with P and m_P its maximal ideal. The map

$$d: \text{Div}(C) \to \mathbb{Z}, \qquad D \to d(D),$$

clearly defines a homomorphism; its kernel is denoted by $\text{Div}_0(C)$ and is called the group of divisors of degree zero.

We shall say that an element $x \in K$ is divisible by the divisor D if $\text{ord}_P(x) \geq \text{ord}_P(D)$ for all closed points P. Two elements x and y in K are said to be congruent modulo D if $x - y$ is divisible by D; this defines an equivalent relation which is usually denoted by $x \equiv y \bmod D$.

Let S be a set of closed points of C and $D \in \text{Div}(C)$; we define a subset of the multiplicative group $K^\times = K - \{0\}$ by

$$\mathscr{L}(D)_S = \{x \in K^\times : \text{ord}_P(x) \geq -\text{ord}_P(D) \text{ for all } P \in S\};$$

it is easily verified that the set

$$L(D)_S = \mathscr{L}(D)_S \cup \{0\}$$

forms a vector space over the field of constants k. We also have the following properties: (i) if $D_1 \geq D_2$, then $\mathscr{L}(D_1)_S \supset \mathscr{L}(D_2)_S$; (ii) if $S \subset S'$, then $\mathscr{L}(D)_{S'} \subset \mathscr{L}(D)_S$; (iii) if $\text{ord}_P D_1 = \text{ord}_P D_2$ for all $P \in S$, then $\mathscr{L}(D_1)_S = \mathscr{L}(D')_S$. For a set S of closed points and a divisor D in $\text{Div}(C)$ we define a new divisor whose support is contained in S by

$$D_S = \sum_P n_P \cdot P,$$

where $n_P = \text{ord}_P(D)$ if $P \in S$ and 0 otherwise.

Theorem 2.1 *Let S be a finite set of closed points. If D, D' are two divisors satisfying $D \leq D'$, then $L(D)_S \subseteq L(D')_S$ and*

$$\dim_k L(D')_S/L(D)_S = d(D'_S) - d(D_S).$$

Proof. Since $\text{ord}_P D \leq \text{ord}_P D'$ for all $P \in S$, the inequality

$$\text{ord}_P(f) + \text{ord}_P(D') = \text{ord}_P(f) + \text{ord}_P(D) + \text{ord}_P(D') - \text{ord}_P(D) \geq 0,$$

holds for all $f \in L(D)_S$ and all $P \in S$; hence $L(D)_S \subseteq L(D')_S$. To establish the second assertion we observe that $L(D)_S = L(D_S)_S$ and $L(D')_S = L(D'_S)_S$; we may thus suppose that $D = D_S$ and $D' = D'_S$. Clearly it suffices to prove that for any closed point Q in S we have

$$\dim_k L(D + Q)_S/L(D)_S = d(Q).$$

In fact, if $D' = D + Q_1 + \cdots + Q_h$, then the inclusions

$$L(D)_S \subseteq L(D + Q_1)_S \subseteq \cdots \subseteq L(D + Q_1 + \cdots + Q_h)_S = L(D')_S$$

yield

$$\dim_k L(D')_S/L(D)_S$$

$$= \sum_{i=1}^{h} \dim_k L(D + Q_1 + \cdots + Q_i)_S/L(D + Q_1 + \cdots + Q_{i-1})_S;$$

the second claim will follow from

$$d(D') - d(D) = \sum_{i=1}^{h} d(Q_i).$$

Let $S = \{P_1, \ldots, P_n\}$ and suppose $P_1 = Q$. Let $k_Q = R_Q/m_Q$ be the residue class field of Q with degree $d = d(Q) = [k_Q : k]$.

Claim $\dim_k L(D + Q)_S/L(D)_S = d$. By the approximation theorem (Theorem 1.3, chapter 1) there exists an element $u \in K^\times$ such that

$$\operatorname{ord}_P(u) + \operatorname{ord}_P(D + Q) = 0 \qquad \text{for all } P \in S.$$

Let $x'_1, \ldots, x'_d$ be elements in the ring R_Q whose images $\bar{x}'_1, \bar{x}'_2, \ldots, \bar{x}'_d$ under the reduction map $R_Q \to R_Q/m_Q = k_Q$ form a basis for k_Q over k. By the approximation theorem (Theorem 1.3, Chapter 1) elements $x_1, \ldots, x_d$ in $K^\times$ can be found such that $\operatorname{ord}_Q(x_j - x'_j) \geq 1$ and $\operatorname{ord}_P(x_j) \geq 0$ for $P \in S - \{Q\}$. Under the map $R_Q \to R_Q/m_Q$ we have $x_j \to \bar{x}_j \in k_Q$. Any element $x \in R_Q$ has a representation of the form

$$x = \sum_{j=1}^{d} a_j x_j + x',$$

with $x' \in m_Q$. Since u satisfies $\operatorname{ord}_P u + \operatorname{ord}_P(D + Q) = 0$ and $\operatorname{ord}_P(x_j) \geq 0$ for all $j = 1, \ldots, d$ and all $P \in S$, we obtain $\operatorname{ord}_P(ux_j) + \operatorname{ord}_P(D + Q) \geq 0$; hence $ux_j \in L(D + Q)_S$. An element $x \in L(D + Q)_S$ satisfies $\operatorname{ord}_Q(X) + \operatorname{ord}_Q(D + Q) \geq 0$ and hence $\operatorname{ord}_Q(xu^{-1}) = \operatorname{ord}_Q(x) - \operatorname{ord}_Q(u) = \operatorname{ord}_Q(u) + \operatorname{ord}_Q(D + Q) \geq 0$, that is to say $xu^{-1} \in R_Q$; therefore we have

$$xu^{-1} = \sum_{j=1}^{d} a_j x_j + x',$$

with $a_j \in k$ and $x' \in m_Q$. Since $\operatorname{ord}_P(xu^{-1}) \geq 0$ and $\operatorname{ord}_P(x_j) \geq 0$ for $j = 1, \ldots, d$ and all $P \in S - \{Q\}$ it follows that $\operatorname{ord}_P(x') \geq 0$ and therefore $\operatorname{ord}_P(ux') + \operatorname{ord}_P(D) = \operatorname{ord}_P(ux') + \operatorname{ord}_P(D + Q) \geq 0$ for all $P \in S - \{Q\}$. For $P = Q$ we have $x' \in m_Q$ and hence $\operatorname{ord}_Q(ux') \geq 1 - \operatorname{ord}_Q(D + Q) \geq -\operatorname{ord}_Q(D)$ or equivalently $\operatorname{ord}_Q(x'u) + \operatorname{ord}_Q(D) \geq 0$; hence $ux' \in L(D)_S$. Thus any $x \in L(D + Q)_S$ is representable in the form

$$x = \sum_{j=1}^{d} a_j x_j u + x'u,$$

with $x'u \in L(D)_S$; hence $\dim_k L(D + Q)_S/L(D)_S \leq d$. The inequality in the other direction will follow if we show that for any non-trivial choice of $a_1, \ldots, a_d$ in k, the element $y = \sum_{j=1}^{d} a_j x_j u$ does not belong to $L(D)_S$. To verify this we observe that the image of yu^{-1} under the map $R_Q \to R_Q/m_Q$ is $\sum_{j=1}^{d} a_j \bar{x}_j' \neq 0$, and hence $\operatorname{ord}_Q(yu^{-1}) = 0$ or equivalently $\operatorname{ord}_Q y + \operatorname{ord}_Q(D + Q) = \operatorname{ord}_Q(y) + \operatorname{ord}_Q(D) + 1 = 0$. On the other hand if $y \in L(D)_S$ then $\operatorname{ord}_Q y + \operatorname{ord}_Q(D) \geq 0$, which is impossible. This completes the proof of Theorem 2.1.

2.2 The vector space $L(D)$

The statement of the Riemann–Roch theorem refers to a vector space which generalizes $L(D)_S$ and whose definition we now present.

Definition 2.2 *Let S be the set of all closed points of C. For D a divisor in* $\operatorname{Div}(C)$ *we put*

$$\mathscr{L}(D) = \{x \in K^\times : \operatorname{ord}_P(x) + \operatorname{ord}_P(D) \geq 0 \text{ for all } P \in S\}$$

and

$$L(D) = \mathscr{L}(D) \cup \{0\}.$$

$L(D)$ is a vector space over k and its dimension is denoted by

$$l(D) = \dim_k L(D).$$

It is easily verified that for two divisors in $\operatorname{Div}(C)$ which satisfy $D \leq D'$ we have $L(D) \subseteq L(D')$.

Theorem 2.2 (i) *If D is any divisor in* $\operatorname{Div}(C)$ *we have that L(D) is of finite dimension over k;*
   (ii) *For any two divisors satisfying* $D \leq D'$ *we have*

$$l(D') - l(D) \leq d(D') - d(D).$$

Proof. We first verify (ii). Let the support of D and D' be contained in the finite set S, so that $L(D) = L(D') \cap L(D)_S$. By the first isomorphism theorem we have

$$L(D')/L(D) = L(D')/L(D') \cap L(D)_S$$
$$\simeq (L(D') + L(D)_S)/L(D)_S \subseteq L(D')_S/L(D)_S;$$

thus

$$\dim_k L(D')/L(D) \le \dim_k L(D')_S/L(D)_S = d(D') - d(D).$$

It remains to show that $L(D')$ is of finite dimension over k.

Let D be a positive divisor different from the zero divisor and satisfying $-D \le D'$. We claim that $L(-D) = \{0\}$. In fact if $x \in L(-D)$, then x cannot be a constant because $\mathrm{ord}_P(x) \ge \mathrm{ord}_P(+D) > 0$ for some closed point P. The function x must be constant for otherwise it would have a pole at some point Q, thus contradicting the assumption that $\mathrm{ord}_P(x) \ge \mathrm{ord}_P(D) \ge 0$ for all P. This shows that $L(-D) = \{0\}$ and hence by (ii) $\dim_k L(D') = d(D') + d(D)$.

Remark If $\mathcal{O}$ denotes the zero divisor, i.e. the divisor with $\mathrm{ord}_P(\mathcal{O}) = 0$ for all P, then $L(\mathcal{O}) = k$ and $l(\mathcal{O}) = 1$.

2.3 Principal divisors and the group of divisor classes

In this section we introduce the notion of a principal divisor and show that its degree is zero. As a consequence we obtain that the set of all principal divisors of degree zero forms a subgroup $\mathrm{Div}_0(C)$ of the group of all divisors. We first need the following lemma.

Lemma 2.1 (i) *If $x \in K^\times$, then $\mathrm{ord}_P(x) = 0$ for all closed points P except a finite number.*

(ii) *The divisor*

$$(x) = \sum_P \mathrm{ord}_P(x)P$$

is well defined.

Proof. Since $\mathrm{ord}_P(x) = 0$ for all $x \in k$, we may assume that x is not a constant. (ii) clearly follows from (i). As for (i), it suffices to show that there is only a finite number of closed points P with $\mathrm{ord}_P(x) > 0$; the same argument will give that $\mathrm{ord}_P(x^{-1}) > 0$ for a finite number of closed points P. Let S be a finite set of closed points for which $\mathrm{ord}_P(x) > 0$. If $\mathcal{O}$ is the zero divisor and $D = \sum'_{P \in S} \mathrm{ord}_P(x)P$, where $\sum'_{P \in S}$ denotes that $\mathrm{ord}_P(x)$ is set equal to zero for all P not in S, then

$$\dim_k L(D)_S/L(\mathcal{O})_S = d(D) = \sum_{P \in S} f_P \mathrm{ord}_P(x).$$

Claim We have

$$\sum_{P \in S} f_P \operatorname{ord}_P(x) \leq [K : k(x)].$$

Suppose $[K : k(x)] = N$ and select $N + 1$ elements $y_0, \ldots, y_N$ in $L(\mathcal{O})_S$ and put

$$\sum_{j=0}^{N} f_j y_j = 0, \qquad f_j \in k[x],$$

where not all the polynomials $f_j(x)$ have zero constant term. Let $f_j(x) = a_j + x g_j(x)$ and rewrite the last equation as

$$\sum_{j=0}^{N} a_j y_j = -x \sum_{j=0}^{N} g_j y_j;$$

from this we obtain

$$\operatorname{ord}_P \left(\sum_{i=0}^{N} a_j y_j \right) = \operatorname{ord}_P(x) + \operatorname{ord}_P \left(\sum_{j=0}^{N} g_j y_j \right) \geq \operatorname{ord}_P(x);$$

this implies that $\sum_{j=0}^{N} a_j y_j$ is an element of $L(-D)_S$. Hence

$$\operatorname{card} S \leq \sum_{P \in S} f_P \operatorname{ord}_P(x) = d(D) - \dim_k L(\mathcal{O})_S / L(-D)_S \leq N.$$

This proves the lemma.

Definition 2.3 *The divisor of zeros of an element $x \in K^\times$ is*

$$(x)_0 = \sum_P n_P P,$$

where $n_P = \operatorname{ord}_P(x)$ if $\operatorname{ord}_P(x) > 0$ and $n_P = 0$ otherwise. The divisor of poles of an element $x \in K^\times$ is

$$(x)_\infty = \sum_P -m_P P,$$

where $m_P = \operatorname{ord}_P(x)$ if $\operatorname{ord}_P(x) < 0$ and $m_P = 0$ otherwise.

Remark The principal divisor of x is

$$(x) = (x)_0 - (x)_\infty = \sum_P \operatorname{ord}_P(x) P,$$

where the sum is taken over all closed points P.

If $x, y \in K^\times$, then $(xy) = (x) + (y)$ and the set

$$\operatorname{Div}_a(C) = \{(x) : x \in K^\times\}$$

of all principal divisors is a subgroup of Div(C). The proof of Lemma 2.1 shows that

$$d((x)_0) \le [K : k(x)] \qquad \text{and} \qquad d((x)_\infty) \le [K : k(x)].$$

Lemma 2.2 *If x is a non-constant element in K^x, then*

$$d((x_0) = d((x)_\infty) = [K : k(x)].$$

Proof. We first show that if $y \in K$ is an integral algebraic function of x, i.e. y satisfies a monic polynomial equation

$$y^m + f_{m-1}(x)y^{m-1} + \cdots + f_0(x) = 0,$$

with $f_j \in k[x]$, then P is a pole of y only if it is a pole of x; equivalently, P appears in $(y)_\infty$ only if it appears in $(x)_\infty$. Observe that if P does not appear in $(x)_\infty$, then $\text{ord}_P(x) \ge 0$ and therefore

$$m \, \text{ord}_P(y) = \text{ord}_P(y^m) = \text{ord}_P\left(\sum_{j=0}^{m-1} f_j(x)y^j\right) \ge \min_{0 \le j \le m-1} \{j \, \text{ord}_P(y), 0\}$$

$$= j_0 \, \text{ord}_P(y),$$

with $0 \le j_0 \le m - 1$. The inequality $(m - j_0)\text{ord}_P(y) \ge 0$ implies that $\text{ord}_P(y) \ge 0$, i.e. P does not appear in $(y)_\infty$. Let us also observe that if y satisfies the equation

$$f_m(x)y^m + f_{m-1}(x)y^{m-1} + \cdots + f_0(x) = 0,$$

with $f_j \in k[x]$, then $z = f_m(x)y$ satisfies the equation

$$z^m + f_{m-1}(x)z^{m-1} + \cdots + f_m(x)^{m-1}f_0(x) = 0$$

and hence z is an integral algebraic function of x. Therefore it is always possible to find a basis of K over $k(x)$ consisting of $N = [K : k(x)]$ integral algebraic elements $y_1, y_2, \ldots, y_N$ of x. The elements $x^i y_j$, $1 \le i \le t, 1 \le j \le N$ are linearly independent over k for any integer t. Since a pole of any y_j is also a pole of x, we have that the divisor $(x^s)_\infty + (y_j) = s(x)_\infty + (y_j)_0 - (y_j)_\infty$ is integral for all sufficiently large positive integers s; choose such an s and fix it. Similarly all the divisors $(x^{s+t})_\infty + (x^i) + (y_j)$, $0 \le i \le t$, $1 \le j \le N$, are integral; this implies that the $N(t + 1)$ elements $x^i y_j$ are linearly independent elements of the vector space $L((x^{s+t})_\infty) = L((s + t)(x)_\infty)$. From Theorem 2.2 we obtain

$$N(t + 1) \le \dim_k L((x^{s+t})_\infty) \le l((x)_\infty) + d((x^{s+t})_\infty) - d((x)_\infty)$$

$$= l((x)_\infty) + (s + t - 1)d((x)_\infty);$$

this implies

$$\frac{N(t+1) - l((x)_\infty)}{s+t-1} \le d((x)_\infty).$$

Letting t go to infinity we obtain that $N \le d((x)_\infty)$; this together with the upper bound obtained earlier shows that $d((x)_\infty) = N$. The same argument but now applied to x^{-1}, using the fact that $k(x) = k(x^{-1})$, gives $d((x)_0) = N$. This completes the proof of the lemma.

Lemma 2.2 states a fundamental property of the function field K, namely the degree of any principal divisor (x) is 0:

$$d((x)) = \sum_P f_P \operatorname{ord}_P(x)$$

$$= d((x)_0) - d((x)_\infty).$$

Remark The relation $d((x)) = 0$ is often referred to as the *product formula* for the function field K.

Another important implication of Lemma 2.2 is that the group $\operatorname{Div}_a(C)$ consisting of all principal divisors of K is a subgroup of $\operatorname{Div}_0(C)$. Inclusion and composition with the degree map yields the sequence

$$0 \to \operatorname{Div}_a(C) \to \operatorname{Div}_0(C) \to \operatorname{Div}(C) \xrightarrow{d} \mathbb{Z} \to 0.$$

The fact that the degree map $\operatorname{Div}(C) \to \mathbb{Z}$ is onto is equivalent to the existence of divisors in C of degree 1. This will be demonstrated in Chapter 3. From the above sequence we obtain

$$0 \to \operatorname{Cl}_0(C) \to \operatorname{Cl}(C) \to \mathbb{Z} \to 0,$$

where

$$\operatorname{Cl}(C) = \operatorname{Div}(C)/\operatorname{Div}_a(C)$$

is the group of *divisor classes* and

$$\operatorname{Cl}_0(C) = \operatorname{Div}_0(C)/\operatorname{Div}_a(C)$$

is the group of *divisor classes of degree* 0.

Remark The group $\operatorname{Cl}_0(C)$ plays an important role in the study of the arithmetic and geometric properties of the curve C. When the field of constants k is algebraically closed, the group $\operatorname{Cl}_0(C)$, which is usually denoted by $\operatorname{Jac}(C)$, has the structure of an algebraic variety with an abelian group law. For instance the classification of all abelian coverings of C, i.e. those whose function fields are abelian extensions of K, are parametrized by the subgroups $\mathrm{Cl}_0(C)$.

From Lemma 2.2 we obtain the following corollary.

Corollary 2.2 *If x is a non-constant element in $K^\times$, then there is an integer μ such that*

$$l((x^m)_\infty) - d((x^m)_\infty) \geq -\mu$$

for all integers m.

Proof. Let t be a positive integer and recall from the proof of Lemma 2.2 the inequality

$$N(t + 1) = \left[K : k\!\left(\frac{1}{x}\right) \right](t + 1) \leq l((x^{t+s})_\infty).$$

If we put $m = s + t$ with $m \geq s$, then

$$l((x^m)_\infty) - d((x^m)_\infty) \geq (1 - s)d((x)_\infty) = -\mu.$$

If $m < s$ we use the fact that $(x^m)_\infty \leq (x^s)_\infty$ together with Theorem 2.2 to obtain

$$l((x^m)_\infty) - d((x^m)_\infty) \geq l((x^s)_\infty) - d((x^s)_\infty) \geq -\mu.$$

This proves the corollary.

2.4 The Riemann theorem

In this section we establish that part of the Riemann–Roch theorem which is originally due to Riemann. This will be used later to prove the full theorem.

Theorem 2.3 (Riemann) *Let x be a non-constant function in K and define an integer g by*

$$1 - g = \min_{m \in \mathbb{Z}} \{l((x^m)_\infty) - d((x^m)_\infty)\}.$$

Then for any divisor D in $\mathrm{Div}(C)$ we have

$$l(D) \geq d(D) + 1 - g;$$

in particular the integer g is independent of x.

Proof. Given a divisor D in $\mathrm{Div}(C)$ we write it as the difference of two positive divisors

$$-D = D_0 - D_\infty.$$

Clearly, since $D_\infty \geq D$ we have

$$l(D) - d(D) \geq l(D_\infty) - d(D_\infty).$$

Hence it suffices to prove the theorem for the divisor D_∞. Let us first observe that the function $l(D) - d(D)$ is constant on the divisor classes, i.e. $l((z) + D) - d((z) + D) = l(D) - d(D)$ for any principal divisor (z) in $\mathrm{Div}_a(C)$ and any divisor D in $\mathrm{Div}(C)$. The equality $d((z) + D) = d((z)) + d(D) = d(D)$ follows from Lemma 2.2 and $l(D) = l((z) + D)$ is a consequence of the isomorphism

$$L(D) \xrightarrow{\sim} L((z) + D)$$

given by $f \to zf$.

Now, since D_∞ is an integral divisor, we have that $(x^m)_\infty \geq -D_\infty + (x^m)_\infty$; by an earlier inequality we have

$$l(-D_\infty + (x^m)_\infty) - d(-D_\infty + (x^m)_\infty) \geq l((x^m)_\infty) - d((x^m)_\infty) \geq 1 - g;$$

this implies

$$l(-D_\infty + (x^m)_\infty) \geq m[K : k(x)] - d(D_\infty) + 1 - g > 0,$$

where the last inequality is certainly true for m sufficiently large. Thus there exists a non-zero element $z \in L(-D_\infty + (x^m)_\infty)$; now $(z) - D_\infty + (x^m)_\infty \geq 0$ implies that $(x^m)_\infty \geq -(z) + D_\infty$ and by Theorem 2.2 this yields

$$\begin{aligned} l(D_\infty) - d(D_\infty) &= l(-(z) + D_\infty) - d(-(z) + D_\infty) \\ &\geq l((x^m)_\infty) - d((x^m)_\infty) \\ &\geq 1 - g. \end{aligned}$$

This proves the theorem.

Definition 2.4 *The genus of the function field K is the integer g defined by*

$$1 - g = \min_D (l(D) - d(D)),$$

where the minimum is taken over all divisors $D \in \mathrm{Div}(C)$.

Remark The definition given above for g is a relative one and depends on the field of constants k. There are situations where the genus of K changes when the field k is replaced by a larger extension. When k is algebraically closed, it is then appropriate to refer to g as the genus of the curve C. In applications to number theory and coding theory, where often $k = \mathbb{F}_q$, the finite field of q elements, the field of K is conservative, i.e. does not change under a finite extensions of k; with these cases in mind we shall simply call g the genus of the curve C.

2. If $\mathcal{O}$ denotes the zero divisor, and if we recall that $l(\mathcal{O}) = 1$, then Riemann's Theorem implies that $1 = l(\mathcal{O}) - d(\mathcal{O}) \geq 1 - g$, i.e. $g \geq 0$.

Definition 2.5 *The degree of speciality of a divisor D is the positive integer*

$$\delta(D) = l(D) - d(D) + g - 1.$$

2.5 Pre-adeles (repartitions)

Let K be the function field of the curve C with k as its field of constants. Let S be the set of all closed points on C.

Definition 2.6 *A family $\{r_P\}_{P \in S}$ of elements $r_P \in K$ is called a pre-adele if*

$$\mathrm{ord}_P(r_P) \geq 0$$

for all $P \in S$ except possibly a finite number.

The element r_P is called the P-th component of the pre-adele $r = \{r_P\}_{P \in S}$. The set A of all pre-adeles is a subset of the set of all maps $P \to r(P)$ from S into K, i.e. a subset of the cartesian product $\prod_{P \in S} K$.

Remark If we were to allow the elements r_P to lie in the completion of K with respect to the valuation ord_P, we would obtain the ordinary notion of adele. As this is not necessary for the proof of the Riemann–Roch Theorem given here, we shall continue to use the concept of pre-adele; in the older literature a pre-adele was called a *repartition*.

If $r = \{r_P\}$ and $r' = \{r_P'\}$ are two pre-adeles, then there are only a finite number of closed points P for which $\mathrm{ord}_P(r_P)$ and $\mathrm{ord}_P(r_P')$ are not both ≥ 0; this allows a definition of addition and multiplication in A, that is, $r + r' = \{(r + r')_P\}_{P \in S}$ and $rr' = \{(rr')_P\}_{P \in S}$, where $(r + r')_P = r_P + r_P'$ and $(rr')_P = r_P \cdot r_P'$. The set of pre-adeles is in fact a subring of the cartesian product $\prod_{P \in S} K$.

With the element $x \in K$ there is associated a (principal) pre-adele $(x_P)_{P \in S}$, where $x_P = x$ for all $P \in S$. It is often convenient to denote the principal pre-adele by the symbol x. The map $K \to A$ given by $x \to x = \{x_P\}_{P \in S}$ is an isomorphism of K into a subring of A. The unit element of K is also the unit element of A; since A contains an isomorphic image of K, we may consider the ring of pre-adeles as a vector space over K and therefore also as a vector space over k. In the following we shall consider A mainly as a vector space over k.

If $r = \{r_P\}$ is a pre-adele and P is a closed point, then we set

$$\operatorname{ord}_P(r) = \operatorname{ord}_P(r_P),$$

where r_P is the P-component of r. Observe that if r is a principal pre-adele, then $\operatorname{ord}_P(r)$ agrees with the earlier definition; we shall refer to $\operatorname{ord}_P(r)$ as the order of the pre-adele r at P. If r and s are two pre-adeles, then from the properties of ord_P we obtain

$$\operatorname{ord}_P(r + s) \geq \min\{\operatorname{ord}_P(r), \operatorname{ord}_P(s)\}, \qquad \operatorname{ord}_P(rs) = \operatorname{ord}_P(r) + \operatorname{ord}_P(s).$$

Moreover if $\operatorname{ord}_P(r) < \operatorname{ord}_P(s)$, we have $\operatorname{ord}_P(r \pm s) = \operatorname{ord}_P(r)$.

Let $D = \sum_P \operatorname{ord}_P(D)P$ be a divisor in $\operatorname{Div}(C)$. Two pre-adeles r and s are said to be congruent modulo D if the inequality $\operatorname{ord}_P(r - s) \geq \operatorname{ord}_P(D)$ holds for all P; this equivalence relation will be denoted by $r \equiv s \bmod D$.

Given a pre-adele $r = \{r_P\}_{P \in S}$ and a divisor $D = \sum_P \operatorname{ord}_P(D)P$, it is of some interest to investigate the existence of a principal pre-adele x such that $x \equiv r \bmod D$. Motivated by this problem and the definition of the spaces $L(D)$, we are led to consider the following set:

$$A(D) = \{r \in A : \operatorname{ord}_P(r) + \operatorname{ord}_P(D) \geq 0 \text{ for all } P\}.$$

Clearly $A(D)$ is a vector subspace of A. By using this notation the above approximation problem can be restated as follows: given an $r \in A$ and $D \in \operatorname{Div}(C)$, is there an $x \in K$ such that $r - x \in A(-D)$? If this is the case, i.e. $r = x + s$ with $s \in A(-D)$, then r belongs to the space $A(-D) + K$ consisting of elements which are sums of elements in $A(-D)$ and K. Conversely, if $r \in A(-D) + K$ then the problem has a solution. We shall now prove that the vector space $A/A(D) + K$ is finite dimensional over k. Observe first that if $D \leq D'$ and $r \in A(D)$, then $\operatorname{ord}_P(r) + \operatorname{ord}_P D' = \operatorname{ord}_P r + \operatorname{ord}_P D + \operatorname{ord}_P D' - \operatorname{ord}_P D \geq 0$ and hence $A(D) \subseteq A(D')$. Let $D \in \operatorname{Div}(C)$ and $r \in A$. Define an integral divisor

$$D_r = \sum_P a(P)P,$$

where $a(P) = 0$ if $\operatorname{ord}_P(r) \geq 0$ and $a(P) = -\operatorname{ord}_P(r)$ otherwise. Clearly $\operatorname{ord}_P(r) + a(P) \geq 0$ and hence $r \in A(D_r)$. Let D' be the g.c.d. of D and D_r, i.e.

$$D' = \sum_P \max(\operatorname{ord}_P(D), \operatorname{ord}_P(D_r))P.$$

Clearly $D_r \leq D'$ and hence $r \in A(D_r) \subset A(D')$. Thus every element of $A/A(D) + K$ belongs to some $(A(D') + K)/(A(D) + K)$. We shall prove below that in fact $(A(D') + K)/(A(D) + K)$ has finite dimension which is bounded by a number which depends only on D.

Lemma 2.3 *If the divisors D, $D' \in \mathrm{Div}(C)$ satisfy $D \leq D'$, then*

$$\dim_k A(D')/A(D) = d(D') - d(D).$$

Proof. Consider the set $S_0 = \{P \in S: \mathrm{ord}_P D \neq 0 \text{ or } \mathrm{ord}_P D' \neq 0\}$ and let

$$L(D')_{S_0} \to A(D)$$

be the map $x \to r_x$ where the P-th component of the pre-adele r_x is $(r_x)_P = x$ if $P \in S_0$ and $(r_x)_P = 0$ if $P \notin S_0$. The map $x \to r_x$ is clearly k-linear. If $x \in L(D)_{S_0}$, then clearly $r_x \in A(D)$. It follows that the map $x \to r_x$ defines an injection.

$$L(D')_{S_0}/L(D)_{S_0} \to A(D')/A(D). \qquad (2.1)$$

We claim that the map (2.1) is onto. Let $r \in A(D')$; by Theorem 1.3 of Chapter 1 on the independence of valuations, there is an element $u \in K^\times$ such that $\mathrm{ord}_P(u - r_P) + \mathrm{ord}_P D \geq 0$ for all $P \in S_0$; also $\mathrm{ord}_P(r_P) + \mathrm{ord}_P(D') \geq 0$. Since $u = (u - r_P) + r_P$, $\mathrm{ord}_P(u) \geq \min\{\mathrm{ord}_P(u - r_P), \mathrm{ord}_P r_P\} \geq \min\{-\mathrm{ord}_P D, -\mathrm{ord}_P D'\} \geq -\mathrm{ord}_P D'$, and hence $u \in L(D')_{S_0}$. Now for appropriate $P \in S_0$ we have $(r_u)_P = u$, and hence $\mathrm{ord}_P((r_u)_P - r_P) \geq -\mathrm{ord}_P D$. If Q is a closed point not in S_0, then $(r_u)_Q = 0$ and therefore $\mathrm{ord}_Q((r_u)_Q - r_Q) = \mathrm{ord}_Q(r_Q) \geq 0$, because $r \in A(D')$. Since those Q not in S_0 appear in D with coefficients equal to zero, we have $\mathrm{ord}_P(r_u - r) + \mathrm{ord}_P D \geq 0$. This proves the map in (2.1) is onto. By Theorem 2.1, the proof of Lemma 2.3 is now complete.

Again suppose that D, $D' \in \mathrm{Div}(C)$ satisfy $D \leq D'$ and observe that the elements $u \in K$ such that $u + A(D) \subseteq A(D')$ are precisely the elements in $K \cap A(D')$, i.e. the principal pre-adeles u such that $\mathrm{ord}_P u + \mathrm{ord}_P D' \geq 0$; but this is precisely the set $L(D')$; thus we have $A(D') \cap (K + A(D)) = A(D) + L(D')$; we also have $L(D') \cap A(D) = L(D)$. This equality and the elementary isomorphism theorems yield

$$V := (A(D') + K)/(A(D) + K)$$

$$\cong A(D')/A(D') \cap (A(D) + K))$$

$$\cong A(D')/(A(D) + L(D'))$$

$$\cong (A(D')/A(D))/((A(D) + L(D'))/A(D))$$

$$\cong (A(D')/A(D))/(L(D')/(L(D') \cap A(D)))$$

$$\cong (A(D')/A(D))/(L(D')/L(D)).$$

By Lemma 2.3 we obtain

$$\dim_k V = d(D') - d(D) - \{l(D') - l(D)\}$$

$$= l(D) - d(D) - \{l(D') - d(D')\}.$$

The numerical function $l(D) - d(D)$ depends only on D and we also know that $l(D') - d(D') \geq 1 - g$, where g is the genus of K. It follows that

$$\dim_k V \leq l(D) - d(D) + g - 1;$$

this bound depends only on the divisor class of D in $\mathrm{Cl}(C)$; it states that the dimension of the vector space is bounded from above by the index of speciality of the divisor class of D. We now prove a more precise result.

Lemma 2.4 *Let A be the ring of all pre-adeles of K. Let D be a divisor in $\mathrm{Div}(C)$ and put $A(D) = \{r \in A: \mathrm{ord}_P r + \mathrm{ord}_P D \geq 0 \text{ for all } P\}$. Then we have*

$$\dim_k A/(A(D) + K) = \delta(D),$$

where $\delta(D) = l(D) - d(D) + g - 1$ is the index of speciality of the divisor class of D.

Proof. We first show that $\dim_k A/(A(D) + K) \leq \delta(D)$. Suppose for a moment that there exist $h > \delta(D)$ elements $r_1, \ldots, r_h$ in A which are linearly independent modulo $A(D) + K$. Then for each i, $1 \leq i \leq h$, there is a divisor D_i' such that $D \leq D_i'$ and $r_i \in A(D_i')$. If this were the case there would then exist a divisor D' which would be a multiple of $D_1', \ldots, D_k'$ with $r_i \in A(D')$, $1 \leq i \leq k$; but this is impossible because $(A(D') + K)/(A(D) + K)$ is contained in $A/(A(D) + K)$ and the latter has dimension bounded from above by $\delta(D)$. To show that the equality $\dim_k A/(A(D) + K) = \delta$ actually holds, we use the fact that there exists a divisor D_0 with $l(D_0) - d(D_0) = 1 - g$; by Riemann's theorem and Theorem 2.2(ii) if D' is another divisor with $D_0 \leq D'$ then $l(D') - d(D') = 1 - g$. Thus if D' is the least common multiple of D and D_0 then

$$\delta(D) = \dim_k (A(D') + K)/(A(D) + K) \leq \dim_k A/(A(D) + K).$$

This completes the proof of Lemma 2.4.

The problem of determining whether a given pre-adele r belongs to $A(D) +$
K can now be formulated as follows: let $\bar{r}$ be the residue class of r modulo
$A(D) + K$. Since $A/A(D) + K$ is a vector space of dimension $\delta(D)$ over k, a
necessary and sufficient condition for $\bar{r}$ to be zero is that $\bar{\omega}(\bar{r}) = 0$ for all
k-linear functionals defined on $A/A(D) + K$. But this can also be formu-
lated by saying that $\omega(r) = 0$ for all linear functionals ω on A which vanish
on all of $A(D) + K$. This reduces the problem of whether $r \in A(D) + K$ to
$\delta(D)$ linearly independent conditions on r. The linear functionals ω are
investigated more closely in the next section.

2.6 Pseudo-differentials (the Riemann–Roch theorem)

This section develops some elementary aspects of the theory of pseudo-
differentials from the point of view of linear functionals on the ring of
pre-adeles whose support is of the form $A(D) + K$; the main result is a
re-interpretation of the index of speciality $\delta(D)$ of a divisor class; this leads
to the fact that the genus g is the dimenison of a certain space of pseudo-
differentials.

Definition 2.7 *A pseudo-differential of K is a linear form*

$$\omega: A/(A(D) + K) \to k$$

for some divisor D.

For notational convenience we let

$$\Omega^s_{K/k}(D) = \text{Hom}_k(A/(A(D) + K), k);$$

we observe that if the divisors D, D' satisfy $D \le D'$, then $\Omega^s_{K/k}(D') \subseteq \Omega^s_{K/k}(D)$.
Hence we can view the space of pseudo-differentials on K as a projective
limit

$$\Omega^s_{K/k} = \varprojlim_{D} \Omega^s_{K/k}(D).$$

Definition 2.8 *Let $\mathcal{O}$ denote the zero divisor of K; the space $\Omega^s_{K/k}(\mathcal{O})$ is called
the space of pseudo-differentials of the first kind.*

Since $\dim_k A/(A(D) + K) = \delta(D)$, by duality we also have

$$\dim_k \Omega^s_{K/k}(D) = \delta(D).$$

In particular, if we recall that for the zero divisor $\mathcal{O}$, $l(\mathcal{O}) = 1$ and $d(\mathcal{O}) = 0$, we have that the space of pseudo-differentials of the first kind is of dimension g, i.e.

$$\dim_k \Omega^s_{K/k}(\mathcal{O}) = g.$$

We now want to show that $\Omega^s_{K/k}$ (also known as the *dualizing module*) may also be viewed as a vector space over K. In fact think of K as the subset of principal pre-adeles in A and for $x \in K$ and $\omega \in \Omega^s_{K/k}(D)$ put

$$x\omega(r) = \omega(xr)$$

for any $r \in A/(A(D) + K)$. It is easily verified that $x\omega$ is a linear function on A, which vanishes on $A(D - (x)) + K$; hence $x\omega \in \Omega^s_{K/k}(D - (x))$. The following properties are immediate consequences of the definitions: for x, $y \in K$ and ω, $\omega' \in \Omega^s_{K/k}$ we have

   (i) $(xy)\omega = x(y\omega)$,
   (ii) $(x + y)\omega = x\omega + y\omega$
   (iii) $x(\omega + \omega') = x\omega + x\omega'$.

This shows that $\Omega^s_{K/k}$ is a vector space over K. We have the following more precise result.

Theorem 2.4 *Let K be the function field of the curve C with field of constants k. If $\Omega^s_{K/k}$ denotes the space of all pseudo-differentials of K, then*

$$\dim_K \Omega^s_{K/k} = 1.$$

Proof. It suffices to show that two non-zero differentials are linearly dependent over K. Let $\omega \in \Omega^s_{K/k}(D)$ and $\omega' \in \Omega^s_{K/k}(D')$; pick a positive divisor E and consider the k-isomorphism into $\Omega^s_{K/k}(E)$ which results from the two inequalities $\mathrm{ord}_P(xr) + \mathrm{ord}_P(D - (x)) \geq 0$ and $\mathrm{ord}_P(x) + \mathrm{ord}_P(E - D) \geq 0$ for $x \in L(E - D)$:

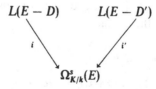

where $i(x) = x\omega$ and $i'(x) = x\omega'$. Suppose the divisor E has been selected with degree $d(E)$ so large that by Riemann's theorem we have the following inequalities:

$$\dim_k\{\text{image of } i\} + \dim_k\{\text{image of } i'\}$$
$$= \dim_k L(E - D) + \dim_k L(E - D')$$

$$\geq d(E - D) + 1 - g + d(E - D') + 1 - g$$

$$\geq l(E) - d(E) + g - 1$$

$$= \dim_k \Omega^s_{K/k}(E).$$

When this is the case then the intersection of the two images is non-empty and hence elements $x, x' \in K$ exist with $x\omega = x'\omega'$. This proves the theorem.

The following two lemmas associate to each pseudo-differential a unique divisor.

Lemma 2.5 *For any two divisors $D, D' \in \mathrm{Div}(C)$, we have*

$$\Omega^s_{K/k}(D) \cap \Omega^s_{K/k}(D') = \Omega^s_{K/k}([D, D']),$$

where $[D, D']$ is the least common multiple of D and D'.

Proof. By definition

$$[D, D'] = \sum_P \max\{\mathrm{ord}_P(D), \mathrm{ord}_P(D')\} \cdot P,$$

and hence $D \leq [D, D']$, $D' \leq [D, D']$; by an earlier remark we have $\Omega^s_{K/k}([D, D']) \subset \Omega^s_{K/k}(D) \cap \Omega^s_{K/k}(D')$. To verify the inclusion in the other direction we consider an element $r \in A([D, D'])$ and observe that

$$\mathrm{ord}_P(r) \geq -\max\{\mathrm{ord}_P(D), \mathrm{ord}_P(D')\}$$

holds for all P. Write $r = r' + r''$ where the P component of r' and r'' are given by $r'_P = r_P$ for all P with $\mathrm{ord}_P D \geq \mathrm{ord}_P D'$ and $r'_P = 0$ and $r''_P = r_P$ for all P with $\mathrm{ord}_P D < \mathrm{ord}_P D'$. Clearly we have $r' \in A(D)$ and $r'' \in A(D')$ (since $\mathrm{ord}_P 0 = \infty$!). This decomposition of r implies $\omega(r) = \omega(r') + \omega(r'') = 0$ for all $\omega \in \Omega^s_{K/k}(D) \cap \Omega^s_{K/k}(D')$. On the other hand, since $\omega(K) = 0$, we obtain that ω vanishes on $A([D, D']) + K$ provided it vanishes on $A(D) + K$ and $A(D') + K$. This proves the lemma.

Lemma 2.6 *We have a well-defined map*

$$\Omega^s_{K/k} \to \mathrm{Div}(C), \qquad \omega \to (\omega),$$

which associates to each non-zero pseudo-differential ω a divisor (ω) in such a way that $\omega \in \Omega^s_{K/k}(D)$ if and only if $(\omega) \leq D$.

Proof. Construction of (ω). Let $\omega \in \Omega^s_{K/k}(D)$ and observe that the map

$$L(D) \to \Omega^s_{K/k}(\mathcal{O}), \qquad x \to x\omega,$$

is a k-isomorphism from $L(D)$ to the space of pseudo-differentials of the first kind $\Omega^s_{K/k}(\mathcal{O})$. By Riemann's theorem we have

$$l(D) \leq \dim_k \Omega^s_{K/k}(\mathcal{O}) = l(\mathcal{O}) - d(\mathcal{O}) + g - 1 = g;$$

from the definition of the index of speciality we have

$$l(D) - d(D) = 1 - g + \delta(D) \geq 1 - g;$$

this implies that if $\Omega^s_{K/k}(D)$ is not empty, then $d(D) \leq 2g - 1$. Now given $\omega \in \Omega^s_{K/k}$ consider all divisors D with $\omega \in \Omega^s_{K/k}(D)$, and from among these choose one of maximal degree and denote it by (ω). Suppose D and D' are two divisors associated with ω. Then, since $\Omega^s_{K/k}([D, D']) = \Omega^s_{K/k}(D) \cap \Omega^s_{K/k}(D')$ and $D \leq [D, D']$ and $D' \leq [D, D']$, we have that $d([D, D']) = \deg D = \deg D'$ and this implies that $D = [D, D'] = D'$, hence the uniqueness of (ω). This proves the lemma.

Corollary 2.5 *If $x \in K$ and $\omega \in \Omega^s_{K/k}$, then $(x\omega) = (x) + (\omega)$.*

Proof. Since $d((x)) = 0$, this is a consequence of the fact that if $\omega \in \Omega^s_{K/k}((\omega))$ and $x \in L(-D)$, then $x\omega \in \Omega^s_{K/k}((\omega) + (x))$.

Corollary 2.5 together with the fact that $\dim_K \Omega^s_{K/k} = 1$ imply that the composite map

$$\Omega^s_{K/k} \to \mathrm{Div}(C) \to \mathrm{Div}(C)/\mathrm{Div}_a(C) = \mathrm{Cl}(C),$$

given by $\omega \to (\omega) \bmod \mathrm{Div}_a(C)$, sends $\Omega^s_{K/k}$ into a unique divisor class.

Definition 2.9 *The canonical class of K is the image of $\Omega^s_{K/k}$ in $\mathrm{Cl}(C)$, i.e. the divisor class of (ω) for some pseudo-differential ω.*

In the following the canonical class will be denoted by W and any divisor in W will be aenoted by w. We are now ready to state the main result of this chapter.

Theorem 2.5 (Riemann–Roch) *If D is a divisor in $\mathrm{Div}(C)$ and w is a divisor in the canonical class W, then*

$$l(D) = d(D) + 1 - g + l(w - D).$$

Proof. We aleary know that $l(D) = d(D) + \delta(D) + 1 - g$ with $\delta(D) = \dim_k \Omega^s_{K/k}(D)$, where $\Omega^s_{K/k}(D)$ is the space of pseudo-differentials vanishing on $A(D) + K$; now, if $\omega \in \Omega^s_{K/k}(w)$, we have an isomorphism $L(w - D) \to \Omega^s_{K/k}(D)$ given by $x \to x\omega$. This proves that $\delta(D) = l(w - D)$. Hence the theorem.

Exercises

1. Let K be a field of algebraic functions of one variable which is separably generated with field of constants k. Show that $K = K^p k(x) = (K^p k)(x)$ for every separating element x of K. In particular, if K is perfect, $K = K^p(x)$ for every separating element x of K.

2. Let $K = k(\mathscr{C})$ be the function field of a proper smooth irreducible curve with exact field of constants k of characteristic p. Show that if $K = k(x, y)$, where y is separable over $k(x)$, then the subfield $k(x^p, y^p)$ is the function field of a curve $\mathscr{C}^{(p)}$ and that the covering map $\pi_p: \mathscr{C} \to \mathscr{C}^{(p)}$ is a purely inseparable rational map of degree p. In particular, show that if k is perfect, in which case $k = k^p$, the map sending (x, y) to (x^q, y^q), for q a fixed power of p, gives an isomorphism of $k(x, y)$ onto $k(x^q, y^q)$ and hence a purely inseparable rational map

$$\pi_q: \mathscr{C} \to \mathscr{C}^{(q)}$$

of degree q.

3. Show that if D is a non-trivial derivation of K (trivial over k), then $Ker(D) = K^p k$. In particular show that if $x \in K$ is a separating element, then every element $y \in K$ can be written uniquely in the form

$$y = \sum_{i=0}^{p-1} u_i x^i$$

with the $u_i \in K^p k$.

4. Let K be a function field of characteristic p. The elements of $K^p k$ are called the *p-constants* of the function field K; all other elements are called *p-variables* of K. Show that every *p-variable* of K is a separating element.

5. Let x and y be *p-variables* and $f(x, y)$ the irreducible equation between them. As $K/K_0(x)$, $K_0 = K^p k$, is separable so is $K_0(y, x)/K_0$, and the formal derivative $f_y \neq 0$; by symmetry we also have $f_x \neq 0$ and the relation $f_x \, dx + f_y \, dy = 0$ holds. Show that if x and y are *p-variables*, and the expansion

$$\frac{dy}{y} = (u_0 + u_1 x + \cdots + u_{p-1} x^{p-1}) \frac{dx}{x}, \, u_i \in K_0,$$

holds, then

$$u_0 = \frac{x \, dy^p}{y \, dx} = \frac{x^p \, dy^p}{y^p \, dx^p},$$

where the differential quotient dy^p/dx^p is to be taken in K^p as the formal derivative of y^p with respect to x^p.

6. (Cartier operation) Let $\Omega_k(K)$ be the space of differentials of the function field K and $\Omega_k(K_0)$ that of $K_0 = K^p k$. Given a differential of the form

$$u \frac{dx}{x} = (u_0 + u_1 x + \cdots + u_{p-1} x^{p-1}) \frac{dx}{x},$$

define a mapping S by setting

$$S\left(u\frac{dx}{x}\right) = u_0\frac{dx^p}{x^p}.$$

Show that the map S is independent of the p-variable x.

7. Show that the Cartier operation and the taking of traces of differentials commute. If the residue of a differential at a point of degree one is defined as usual in terms of the coefficient of T^{-1} in the formal power series expansion of the differential in powers of the local uniformizing parameter T, and for a point of arbitrary degree by taking traces from the residue class field down to the field of constants, show that the sum of the residues of a differential is 0. (*Hints*: (i) Given a fixed differential, make a constant field extension so that all the poles occur at points of degree one. (ii) Prove the statement for the rational function field over an algebraically closed field of constants. (iii) Realize the function field as a separable extension of the rational function field and use the commutativity of the Cartier operation and the taking of traces to deduce the claim.)

Notes

There are many proofs of the Riemann–Roch theorem and almost all of them, with appropriate changes, can be made to work over a field of characteristic p. In applications to questions about linear systems, it is necessary to have a version of the theorem which works for curves with singularities. For a brief statement see R. Hartshorne, *Algebraic Geometry*, page 298, ex. 1.9). See also J. P. Serre, *Groupes Algebriques et Corps des Classes*, Hermann, Paris, 1959. The effective constructibility of basis for the various spaces which appear in the statement of the Riemann-Roch theorem has been pursued by several authors, notably by J. H. Davenport. The original work in this area was done by A. Baker and J. Coates (J. Coates, Construction of rational functions on a curve, *Proc. Cam. Phil. Soc.*, **68**, 1970). The proof of the residue theorem suggested in the exercises is essentially due to Cartier (M. Eichler, *Introduction to the Theory of Algebraic Number and Functions*, Academic Press, New York, 1966). A conceptual approach to the residue theorem based on some ideas of Grothendieck is given in (J. Tate, Residues of differentials on curves, *Ann. Sc. Ec. Normal Sup.*, **4**, 1968).

3

Zeta functions

3.1 Introduction

In this chapter the zeta- and L-functions of algebraic curves over finite fields
are defined and their basic properties developed. The main goal is a
presentation of Bombieri's proof [8] of the Riemann hypothesis for these
functions. Weil's original proof [95] depends on some deep theorems of
algebraic geometry whose study still remains a formidable task for the
beginning student. The proof given in Section 3.4 is based entirely on
the Riemann–Roch theorem and on some elementary Galois theory and
uses a fundamental idea of Stepanov. As Manin has remarked [54, p. 720],
an interesting aspect of Bombieri's proof is its code theoretic nature; some
important consequences of this will be given in Chapter 5.

3.2 The zeta functions of curves

Let $k = \mathbb{F}_q$ be the finite field of q elements; let C be a smooth projective
curve, geometrically connected, and defined over k. Let K be the function
field of C and k its exact field of constants. If P is a closed point of C
and R_P (resp. m_P) its valuation ring (resp. maximal ideal), we define its
degree by

$$d(P) = [k_P : k],$$

where k_P denotes the residue class field R_P/m_P; similarly the norm of P is
defined by

$$N(P) = q^{d(P)};$$

this is also the number of elements in k_P. For a divisor $D = \sum_P \mathrm{ord}_P(D)P$
we put, as in Chapter 2,

$$d(D) = \sum_P \mathrm{ord}_P(D)d(P) \qquad \text{and} \qquad N(D) = q^{d(D)}.$$

Definition 3.1 *The zeta function of the curve $C/\mathbb{F}_q$ is*

$$Z(t) = \sum_D t^{d(D)},$$

where the sum is taken over all the positive divisors D in $\text{Div}(C)$.

A well-known argument of Euler, together with the fact that a positive divisor in $\text{Div}(C)$ is a finite linear combination with positive coefficients of closed points on C, yields the Euler product identity

$$\sum_D t^{d(D)} = \prod_P (1 - t^{d(P)})^{-1},$$

where the product is taken over all closed points P on C, i.e. over all prime divisors P in $\text{Div}(C)$. With the substitution $t = q^{-s}$, the above identity becomes

$$\sum_D N(D)^{-s} = \prod_P (1 - N(P)^{-s})^{-1},$$

which is closer to the definition of the ordinary Riemann zeta function $\zeta(s)$.

We need to show first that $Z(t)$ has a positive radius of convergence; this will be accomplished by showing that $Z(t)$ is in fact a rational function of t.

To a divisor class $\mathcal{D}$ in $\text{Div}(C)$ we associate two numbers:

$n(\mathcal{D})$ = number of positive divisors in $\mathcal{D}$,
$N(\mathcal{D})$ = maximal number of linearly independent positive divisors in $\mathcal{D}$.

Recall that the divisors $D + (f_1), \ldots, D + (f_m)$ in the class $\mathcal{D}$ are said to be linearly independent if the functions $f_1, \ldots, f_m$ are linearly independent over k. If D is a divisor in $\mathcal{D}$, then $N(\mathcal{D}) = l(D)$.

Lemma 3.1 *Let D be a divisor in* $\text{Div}(C)$. *Then the number of positive divisors in* $\text{Div}(C)$ *equivalent to D is* $(q^{l(D)} - 1)/(q - 1)$.

Proof. As in Chapter 2, we consider the set $\mathscr{L}(D) = \{ f \in K^{\times} : (f) + D \geq 0 \}$. The set $L(D) = \mathscr{L}(D) \cup \{0\}$ is a vector space over k of dimension $l(D)$, containing $q^{l(D)}$ elements; the set $\mathscr{L}(D)$ itself contains $q^{l(D)} - 1$ elements. Observe that to each function $f \in \mathscr{L}(D)$ there corresponds a divisor $D' = D + (f)$ which is positive and in the same divisor class as D. Conversely, if D' is such a divisor, then there is a function $f \in \mathscr{L}(D)$ such that $(f) = D' - D$; furthermore, the only functions in $\mathscr{L}(D)$ which have this property are the constant multiples cf of f with $c \in k - \{0\}$ and these are $q - 1$ in number. In fact if $D' = D + (f) = D + (f_1)$, then $(f) = (f_1)$ or equivalently $(f/f_1) = \mathcal{O}$, the zero divisor; this implies that f/f_1 is a constant whose value must necessarily be a non-zero element in k. The relation $(f) = D' - D$ therefore establishes a correspondence between the positive divisors D' in $\text{Div}(C)$ and the functions $f \in \mathscr{L}(D)$, where to each function f there corresponds one divisor D' and to every divisor there corresponds $q - 1$ functions f. Since the number of elements in $\mathscr{L}(D)$ is $q^{l(D)} - 1$, this proves the lemma.

On the curve C we have three distinguished groups of divisors rational over k: (i) $\text{Div}(C)$, the group of all divisors on C generated by the closed points, (ii) $\text{Div}_0(C)$, the subgroup consisting of divisors of degree 0 and (iii) $\text{Div}_a(C)$, the subgroup of principal divisors, i.e. those of the form (f) with $f \in K^\times$. The set $\{d(D): D \in \text{Div}(C)\}$ is a subgroup of $\mathbb{Z}$ of the form $\delta\mathbb{Z}$ with δ a positive integer. If we let $\text{Cl}(C) = \text{Div}(C)/\text{Div}_a(C)$ and $\text{Cl}_0(C) = \text{Div}_0(C)/\text{Div}_a(C)$ then we obtain the exact sequence

$$0 \to \text{Cl}_0(C) \to \text{Cl}(C) \to \delta\mathbb{Z} \to 0,$$

which is of fundamental importance in the study of the zeta function of the curve C. It will be shown later that indeed $\delta = 1$.

Lemma 3.2 *For any integer d, the number of divisor classes in $\text{Cl}(C)$ of degree d is independent of d and is equal to the cardinality of $\text{Cl}_0(C)$.*

Proof. We first show that for a fixed positive integer d_0 there can be at most a finite number of closed points P on C with bounded degree $d(P) \le d_0$. Let x be a non-constant function in $K^\times$ and let $D = (x)_\infty$ be its divisor of poles. Since D is a finite linear combination of closed points, it is enough to show that outside the support of D there can be at most a finite number of closed points Q with $d(Q) \le d_0$. In fact if $Q \notin \text{supp}(D)$, then $\text{ord}_Q(x) \ge 0$ and hence x is an element of the local ring R_Q of Q. The function $P \to X(P)$ defines a covering of the affine line of degree $[K : k(x)]$:

$$
\begin{array}{cc}
Q, & C - \{\text{supp } D\} \\
\vdots & \downarrow \\
\downarrow & \\
P, & \mathbb{P}^1 - \{\infty\} = \mathbb{A}^1 ;
\end{array}
$$

Figure 3.1

the closed point Q lies above a unique closed point P in $\mathbb{A}^1$ which corresponds to a monic irreducible polynomial $p(x)$ in the polynomial ring $A = k[x]$ and we also have $\deg(P) = \deg(p(x)) \leq \deg(Q)$; hence if $d(Q) \leq d_0$, then $d(P) \leq d_0$. Now the number of irreducible polynomials in A of degree $\leq d_0$ is finite; also the number of closed points Q in C above a given closed point P in $\mathbb{A}^1$ is at most $[K : k(x)]$. Therefore the total number of closed points Q in C with $d(Q) \leq d_0$ is finite. To complete the proof of the lemma we assume that the given fixed degree d satisfies $d + 1 > g$ and pick a divisor class $\mathscr{D}$ with $d(\mathscr{D}) = d$. We note that $d \in \delta\mathbb{Z}$. If D is a divisor in $\mathscr{D}$, we have by an appeal to Riemann's theorem, Section 2.4, $l(D) \geq d(D) \geq 1 - g$. Pick a divisor D in $\mathscr{D}$ which is positive. Now any other divisor in $\mathscr{D}$ has the form $D + (f)$ with some $f \in K^\times$; i.e. the class $\mathscr{D}$ can be generated by D. The question that now arises is that of counting the number of positive divisors D which can generate divisor classes of degree d. Observe that if D is such a divisor and P is a closed point contained in the support of D, then

$$d(P) \leq \sum_Q \operatorname{ord}_Q(D) \cdot d(Q) = d(D) = d.$$

By what was proved above we know that the number of such closed points is finite. This then proves that D can only be one of finitely many divisors; hence the number of divisor classes of degree $d > g - 1$ is finite.

To prove that the number of divisor classes of degree d is independent of d we proceed as follows. Select integers d and d' in $\delta\mathbb{Z}$ and pick divisor classes in $\mathscr{D}$ and $\mathscr{D}'$ with degrees d and d'. Suppose there are m divisor classes of degree d, say $S = \{\mathscr{D}_1, \ldots, \mathscr{D}_m\}$, and m' of degree d', say $S' = \{\mathscr{D}'_1, \ldots, \mathscr{D}'_{m'}\}$. Now, in the group $\operatorname{Div}(C)$, the classes $\mathscr{D}_i + \mathscr{D}' - \mathscr{D}$, with $\mathscr{D}_i \in S$, are all distinct and each has degree d', hence they are contained in S' and therefore $m \leq m'$. Since the argument is symmetric in $\mathscr{D}$ and $\mathscr{D}'$, we have $m = m'$. To deal with the divisor classes of degree $< g - 1$ we can first translate them by a divisor with positive degree and then apply the previous argument. In particular, the number of divisor classes of any degree is equal to the number of divisor classes of degree 0, which is simply the cardinality of $\operatorname{Cl}_0(C)$. This proves the lemma.

Remark $\operatorname{Cl}_0(C)$ is the analogue of the class group of an algebraic number field; thus when convenient we shall refer to the cardinality of $\operatorname{Cl}_0(C)$ as the class number of the curve C and denote it by $h(C)$. It should be kept in mind that the value of h depends on the field $k = \mathbb{F}_q$ of definition.

We are now in a position to prove that the zeta function $Z(t)$ converges in some neighborhood of the origin. By definition, we have

$$Z(t) = \sum_D t^{d(D)}$$

$$= \sum_d \sum_{\mathscr{D}} \sum_D t^d,$$

where the sum $\sum_d$ is taken over all non-negative integers in $\delta\mathbb{Z}$, the sum $\sum_{\mathscr{D}}$ is taken over the divisor classes of degree d and the sum $\sum_D$ is taken over all integral divisors in the class $\mathscr{D}$; by Lemmas 3.1 and 3.2 we can rewrite the last expression as

$$= h(C) \sum_d \frac{q^{l(D)} - 1}{q - 1} \cdot t^d,$$

where for each d, D is a positive divisor in a class $\mathscr{D}$ of degree d. Now recall that if D is a positive divisor in a class $\mathscr{D}$ and $l(D) > 0$, then by Theorem 2.2 in Chapter 2, $l(D) \le d(D) + l(\mathcal{O}) = d(\mathscr{D}) + 1$. Hence

$$Z(t) \le h(C) \sum_d (d + 1)(qt)^d;$$

this shows that $Z(t)$ is absolutely convergent inside the circle $q|t| < 1$. Below we will see that $t = q^{-1}$ is actually a simple pole.

The same argument given above applies to the Euler product identity and hence if we substitute $t = q^{-s}$ we obtain that both the Dirichlet series and the Euler product

$$Z(q^{-s}) = \sum_D N(D)^{-s}$$

$$= \prod_P (1 - N(P)^{-s})^{-1}$$

converge absolutely in the half plane $Re(s) > 1$; in this same region $Z(q^{-s})$ is free of zeros.

3.3 The functional equation

In the following we make the substitution $t = q^{-s}$ and write

$$\zeta(s, C) = Z(q^{-s}).$$

In order to add to the formal analogies between $\zeta(s, C)$ and the ordinary Riemann zeta function $\zeta(s)$, we now establish a functional equation which relates its value at s with that at $1 - s$. The argument, which is based on the Riemann–Roch theorem, actually proves more, namely, $\zeta(s, C)$ is a rational function in $\mathbb{Z}(q^{-s})$.

Theorem 3.1 (Functional equation) $\zeta(s, C)$ *is a rational function in* $\mathbb{Z}(q^{-s})$ *and satisfies*

$$\zeta(1 - s, C) = N(W)^{s-1/2}\zeta(s, C),$$

where $N(W) = q^{2g-2}$ is the norm of the canonical class.

Proof. In the infinite series which defines $\zeta(s, C) = Z(q^{-s})$, we first sum over all the possible degrees and then for each such d, we distribute the positive divisors into classes. By Lemma 3.1, we then obtain

$$(q - 1)\zeta(s, C) = \sum_d \sum_{\mathcal{D}} (q^{N(\mathcal{D})} - 1)q^{-ds},$$

where d is taken over all non-negative integers in $\delta\mathbb{Z}$ and for a fixed d, $\mathcal{D}$ runs over the divisor classes of degree d. Here we have used the fact that $N(\mathcal{D}) = l(\mathcal{D})$ for any integral divisor in $\mathcal{D}$. By Lemma 3.2 we obtain

$$(q - 1)\zeta(s, C) = \sum_d \sum_{\mathcal{D}} q^{N(\mathcal{D})-ds} - \frac{h(C)}{1 - q^{-s}}.$$

We now break the last expression into two series,

$$(q - 1)\zeta(s, C) = F(q^{-s}) + G(q^{-s}),$$

$$F(q^{-s}) = \sum_d{}' \sum_{\mathcal{D}} q^{N(\mathcal{D})-ds} - \frac{h(C)}{1 - q^{-s}}$$

and

$$G(q^{-s}) = \sum_d{}'' \sum_{\mathcal{D}} q^{N(\mathcal{D})-ds},$$

where the sum $\sum'$ is taken over those degrees $d \in \delta\mathbb{Z}$ with $d \geq 2g - 2 + \delta$ and $\sum''$ is taken over those degrees $d \in \delta\mathbb{Z}$ with $0 \leq d \leq 2g - 2$. Making the substitution $t = q^{-s}$, we can also write this in the form

$$(q - 1)Z(t) = F(t) + G(t).$$

To evaluate $F(t)$ we use the Riemann–Roch theorem and the fact that if $\mathcal{D}$ is a divisor class of degree $d(\mathcal{D}) > 2g - 2$, then $N(\mathcal{D}) = d(\mathcal{D}) - g + 1$; hence using Lemma 3.2, we can write

$$F(t) = h(C) \sum_d{}' q^{d-g+1}t^d - \frac{h(C)}{1 - t^{\delta}}$$

$$= h(C)q^{1-g}\frac{(qt)^{2g-2+\delta}}{1 - (qt)^{\delta}} - \frac{h(C)}{1 - t^{\delta}}.$$

We have used here implicitly the fact that $2g - 2$ is the degree of the canonical class and as such is a multiple of δ. Since $G(t)$ is a polynomial in t, it is now clear from the last expression that $Z(t)$ is a rational function in t. By a straightforward calculation one easily checks that

$$F(t) = q^{g-1}t^{2g-2}F\left(\frac{1}{tq}\right).$$

Let us now consider the polynomial

$$G(t) = \sum_{\mathscr{D}} q^{N(\mathscr{D})}t^{d(\mathscr{D})},$$

where $\mathscr{D}$ runs over all divisor classes with $0 \le d(\mathscr{D}) \le 2g - 2$. Substituting the value $N(\mathscr{D})$, which by the Riemann–Roch theorem is

$$N(\mathscr{D}) = d(\mathscr{D}) - g + 1 + N(W - \mathscr{D}),$$

into the last sum, we obtain

$$G(t) = \sum_{\mathscr{D}} q^{d(\mathscr{D})-g+1+N(W-\mathscr{D})}t^{d(\mathscr{D})}$$

$$= \sum_{\mathscr{D}} (qt)^{d(\mathscr{D})}q^{1-g}q^{N(W-\mathscr{D})}$$

$$= q^{1-g}(qt)^{2g-2}\sum_{\mathscr{D}} (1/qt)^{d(W-\mathscr{D})}q^{N(W-\mathscr{D})}.$$

Now as $\mathscr{D}$ ranges over all divisor classes with degrees in the range $0 \le d(\mathscr{D}) \le 2g - 2$, so do all the divisor classes $W - \mathscr{D}$. Hence in the last sum we may make the substitution $W - \mathscr{D} \to \mathscr{D}$, thus obtaining

$$G(t) = q^{g-1}t^{2g-2}\sum_{\mathscr{D}} (1/qt)^{d(\mathscr{D})}q^{N(\mathscr{D})}$$

$$= q^{g-1}t^{2g-2}G(1/qt).$$

Putting together the relations for F and G we obtain

$$(q - 1)Z(t) = q^{g-1}t^{2g-2}(q - 1)Z(1/qt);$$

substituting $t = q^{-s}$ and putting $N(W) = q^{d(W)}$, we obtain

$$\zeta(s, C) = N(W)^{1/2-s}\zeta(1 - s, C),$$

which proves the theorem.

The following result is a simple consequence of the above proof.

Corollary 3.1.1 *We have*

$$(q - 1)Z(t) = \sum_{\mathscr{D}} q^{N(\mathscr{D})}t^{d(\mathscr{D})} + h(C)q^{1-g}\frac{(qt)^{2g-2+\delta}}{1 - (qt)^{\delta}} - \frac{h(C)}{1 - t^{\delta}},$$

where the sum $\sum_{\mathscr{D}}$ is taken over all divisor classes with degrees in the range $0 \le d(\mathscr{D}) \le 2g - 2$.

The rational representation for $Z(t)$ which appears in the above corollary implies that the poles of $Z(t)$ correspond to the roots of the equations

$$1 - (qt)^\delta = 0 \quad \text{and} \quad 1 - t^\delta = 0.$$

When we establish that $\delta = 1$, it will result that the only poles of $Z(t)$ are $t = 1$ and $t = 1/q$ and these are simple.

The following result, which will be useful in the proof of the Riemann hypothesis for $\zeta(s, C)$ in Section 3.4, describes the behavior of the zeta function of the curve C when the base field $\mathbb{F}_q$ undergoes finite algebraic extensions.

Lemma 3.3 *Let C_n be the curve obtained from C by extending the field of constants from $\mathbb{F}_q$ to $\mathbb{F}_{q^n}$. We then have*

$$Z(t^n, C_n) = \prod_\zeta Z(\zeta t, C),$$

where the product is taken over all the n-th roots of 1. Here $Z(T, C)$ is the zeta function of the curve C and $Z(T, C_n)$ that of C_n.

Proof. The basic idea of the proof is to compare the Euler products which define $Z(t, C)$ and $Z(t, C_n)$. To this effect we consider the covering $C_n \to C$ and investigate how a closed point P in C splits into closed points Q in C_n.

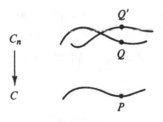

Figure 3.2

Pick a closed point P of C and let R_P, and m_P be its local ring and its maximal ideal respectively. The residue field $\mathbb{F}_P = R_P/m_P$ is a finite extension of $\mathbb{F}_q$ of degree $d(P)$. We now select an element $x \in R_P$ such that its image $\bar{x}$ generates $\mathbb{F}_P$, i.e. $\mathbb{F}_P = \mathbb{F}_q(\bar{x})$. Consider an irreducible polynomial $f(T) \in \mathbb{F}_q[T]$ of degree $d(P)$ and satisfying $f(\bar{x}) = 0$; for such a polynomial we have

$$\mathbb{F}_P \cong \mathbb{F}_q[T]/(f).$$

Now, since $f(T)$ is defined over $\mathbb{F}_q$ and irreducible, we may assume it is separable. Let us consider the factorization of $f(T)$ in $\mathbb{F}_{q^n}$. By the theory of finite fields, we know that in $\mathbb{F}_{q^n}$, $f(T)$ splits as a product

$$f(T) = f_1(T) \dots f_e(T),$$

of distinct irreducible polynomials of the same degree, where e is the greatest common divisor of n and $d(P)$. This factorization of $f(T)$ gives in turn a splitting of the divisor corresponding to P into a divisor

$$P = Q_1 + \cdots + Q_e,$$

when considered as a divisor in C_n, where each Q_i is a closed point in C_n. Observing that the degree of a divisor remains invariant under extensions of the base field, we must have

$$d(P) = d(Q_1) + \cdots + d(Q_e) = ed(Q_1),$$

and therefore $d(Q_i) = d(P)/e$ for $1 \le i \le e$. We have thus shown that to each closed point P in C, there corresponds e closed points in C_n each of degree $d(P)/e$. Hence for each closed point P in C, which contributes a local factor in $Z(t, C)$, the Euler product $Z(t^n, C_n)$ contains e factors of the form $(1 - t^{nd(P)/e})$. The simple identity

$$(1 - t^{nd(P)/e})^e = \prod_\zeta (1 - (\zeta t)^{d(P)}),$$

where the product runs over all the n-th roots of 1, can be obtained by a formal manipulation from the Maclaurin series for $\log(1 - T)$; using this identity between the local factors of $Z(t^n, C_n)$ and $Z(t, C)$ we obtain, by taking the product over all closed points P in C,

$$Z(t^n, C_n) = \prod_\zeta Z(\zeta t, C).$$

This proves the lemma.

If we combine the identities of Lemma 3.3 and the Corollary 3.1.1, with $n = \delta$, we obtain

$$Z(t^\delta, C_\delta) = \frac{L(t^\delta)}{(1 - t^\delta)^\delta (1 - (qt)^\delta)^\delta},$$

where $L(T)$ is a polynomial in $\mathbb{Z}[T]$ with $L(0) = 1$. Since we already know that the poles of $Z(t^\delta, C_\delta)$ are simple, it must be true that $\delta = 1$. Hence we have the following result.

Theorem 3.2 *Let C be a complete non-singular curve defined over $\mathbb{F}_q$ with function field K; suppose that $\mathbb{F}_q$ is the exact field of constants of K. We then* *have that*

(i) *The degree map induces an exact sequence*

$$0 \longrightarrow \mathrm{Cl}_0(C) \longrightarrow \mathrm{Cl}(C) \xrightarrow{\deg} \mathbb{Z} \longrightarrow 0,$$

where $\mathrm{Cl}(C)$ *is the group of divisor classes of* C *and* $\mathrm{Cl}_0(C)$ *the subgroup of those of degree 0,*

(ii) *The zeta function of* C *can be represented in the form*

$$Z(t, C) = \frac{L(t)}{(1-t)(1-qt)}$$

where $L(t)$ *is a polynomial in* $\mathbb{Z}[t]$ *of degree* $2g$ *with* $L(0) = 1$ *satisfying*

$$L(t) = q^g t^{2g} L(1/qt),$$

(iii) *the residue of* $Z(t, C)$ *at* $t = 1$ *is* $h(C) = L(1)$, *i.e.* $L(1)$ *is the number of divisor classes of* C *of degree* 0.

Remarks 1. The first assertion of the theorem can also be stated as saying that on a complete non-singular curve C defined over $\mathbb{F}_q$, there is always a divisor of degree 1.

2. If $C = \mathbb{P}^1$ the projective line, then

$$Z(t, \mathbb{P}^1) = \frac{1}{(1-t)(1-qt)};$$

in particular for any curve C we have

$$\frac{Z(t, C)}{Z(t, \mathbb{P}^1)} = L(t),$$

which is a polynomial of degree $2g$.

3.3.1 Consequences of the functional equation

Let C be a non-singular complete curve defined over $\mathbb{F}_q$; taking the logarithmic derivative of both sides of the Euler product

$$Z(t, C) = \prod_P (1 - t^{d(P)})^{-1}$$

and multiplying by t we obtain

$$t\frac{Z'}{Z}(t, C) = \sum_{n=1}^{\infty} \sum_P d(P) t^{nd(P)}$$

$$= \sum_{m=1}^{\infty} \left(\sum_P d(P) \right) t^m,$$

where the sum $\sum_P$ is taken over those closed points rational over $\mathbb{F}_q$ whose degree $d(P)$ divides m. The last expression can also be written in the more suggestive form

$$Z(t, C) = \exp\left\{\sum_{m=1}^{\infty} N_m \frac{t^m}{m}\right\},$$

where

$$N_m = \sum_P d(P), \qquad (d(P)|m).$$

Clearly when $m = 1$, the sum $\sum_P 1$ simply counts the number of closed points of C rational over $\mathbb{F}_q$, i.e. those prime divisors of degree 1; this is equivalent to saying that in the exponential representation of $Z(t, C)$, the coefficient of t is equal to the number of closed points on C which are rational over $\mathbb{F}_q$. A similar interpretation can be given for the coefficients N_m of the higher powers t^m. In fact the identity given in Lemma 3.3 can be written in the form

$$Z(t^n, C_n) = \exp\left\{\sum_{m=1}^{\infty} M_m \frac{t^{nm}}{m}\right\}$$

$$= \exp\left\{\sum_{m=1}^{\infty} N_m \frac{t^m}{m}\left(\sum_{\zeta} \zeta^m\right)\right\},$$

where C_n is the curve obtained from C by extending the field of constants from $\mathbb{F}_q$ to $\mathbb{F}_{q^n}$. If we recall that

$$\sum_{\zeta} \zeta^m = \begin{cases} n & \text{if } n/m \\ 0 & \text{otherwise,} \end{cases}$$

we obtain, by comparing the coefficients of t^n on both sides, that $N_n = N_1(C_n)$; in other words N_n is the number of closed points on C which are rational over $\mathbb{F}_{q^n}$, i.e. the prime divisors whose degree divides n.

The fact that the polynomial $L(t)$ which appears in the rational representation

$$Z(t, C) = \frac{L(t)}{(1 - t)(1 - tq)}$$

has the value $L(0) = 1$ and satisfies the functional equation

$$L(t) = q^g t^{2g} L(1/qt),$$

together with the formal expansion

$$L(t) = (1 - t)(1 - qt)Z(t)$$

$$= (1 - t)(1 - qt)\exp\left\{\sum_{n=1}^{\infty} N_n \frac{t^n}{n}\right\}$$

$$= 1 + (N_1 - q - 1)t + O(t^2),$$

valid for small t, imply that

$$L(t) = 1 + (N_1 - q - 1)t + \cdots + q^g t^{2g}.$$

If we let $\alpha_1, \ldots, \alpha_{2g}$ be the reciprocals of the roots of $L(t)$ and put

$$Z(t, C) = \frac{\prod_{i=1}^{2g}(1 - \alpha_i t)}{(1 - t)(1 - qt)},$$

then, after exponentiating the right-hand side, we obtain

$$\exp\left\{\sum_{n=1}^{\infty} N_n \frac{t^n}{n}\right\} = \exp\left\{\sum_{n=1}^{\infty}\left(q^n + 1 - \sum_{i=1}^{2g}\alpha_i^n\right)\frac{t^n}{n}\right\};$$

comparing the coefficients of t^n above we get the identity

$$N_n = q^n + 1 - \sum_{i=1}^{2g}\alpha_i^n,$$

which will play a fundamental role in the proof of the Riemann hypothesis for $Z(t, C)$ given in Section 3.4. The following result is also an immediate consequence of the functional equation for $Z(t, C)$.

Corollary 3.2.1 *The reciprocals* $\alpha_1, \ldots, \alpha_{2g}$ *of the roots of* $L(t)$ *can be numbered so that*

$$\alpha_i \alpha_{g+i} = q, \qquad 1 \leq i \leq g.$$

3.4 The Riemann Hypothesis

Let C be a complete non-singular curve defined over $\mathbb{F}_q$, and let $\zeta(s, C) = Z(q^{-s}, C)$ be its zeta function. By the Riemann hypothesis, henceforth referred simply as RH, for the curve C we understand the statement that all the zeros of $\zeta(s, C)$ are located on the critical line $Re(s) = \frac{1}{2}$. It is immediate that this statement is equivalent to the following theorem.

Theorem 3.3 (Riemann hypothesis) *The reciprocal roots* $\alpha_1, \ldots, \alpha_{2g}$ *of the numerator of*

$$Z(t, C) = \frac{L(t)}{(1-t)(1-qt)}$$

all satisfy $|\alpha_i| = q^{1/2}$, $1 \le i \le 2g$.

The number theoretic importance of this result lies in its many diophantine implications of which the most immediate is

$$|N_n - q^n - 1| = |\alpha_1^n + \cdots + \alpha_{2g}^n| \le 2gq^{n/2}. \tag{3.1}$$

As a consequence of the formula which appears in Lemma 3.3 we make a remark which will be used repeatedly.

Corollary 3.3.1 *If C_n is the curve obtained from C by extending the field of constants from $\mathbb{F}_q$ to $\mathbb{F}_{q^n}$, then the Riemann hypothesis holds for $Z(t, C_n)$ if and only if it holds for $Z(t, C)$.*

The claim in one direction is clear since $Z(t, C)$ is a factor of $Z(t^n, C_n)$. To show that if the RH holds for $Z(t, C)$ then it also holds for $Z(t, C_n)$ we use the diophantine statement (3.1) and the observation already made in Section 3.3.1 to the effect that N_n also represents the number of closed points on $C_n/\mathbb{F}_{q^n}$ which are of degree 1 over $\mathbb{F}_{q^n}$. We leave the details as a simple exercise.

Remark The above corollary will be used in the course of the proof of the RH to justify several extensions of the field of constants.

As already indicated in the introduction, the proof of the RH given here is due to Bombieri; it depends on two basic ideas; the first of these develops an idea of Stepanov and consists of the construction of a function with a prescribed number of zeros at a given finite set of points. The second idea, which is perhaps more subtle, consists of an ingenious use of Galois theory to convert upper bound estimates into lower bounds.

In the following we break the proof of the RH into several steps of which the first is the following.

Lemma 3.4 *The following statements are equivalent.*

(i) *The RH holds for $Z(t, C)$.*

(ii) *There exists constants A, B and a positive integer N such that*

$$|N_d - (q^d + 1)| \le A + Bq^{d/2}$$

holds for all sufficiently large multiples d of N.

Proof. We have already remarked that (i) implies (ii). Hence we assume that (ii) holds and try to deduce the RH. By Corollary 3.3.1 above we may replace the field of constants $\mathbb{F}_q$ by $\mathbb{F}_{q^n}$ in the proof of (i); therefore we may assume that the inequality

$$\left| \sum_{j=1}^{2g} \alpha_j^d \right| \le A + Bq^{d/2}$$

already holds for all $d \ge 1$. Also, since $\prod_{j=1}^{2g} \alpha_j = q^g$, it suffices to prove that $|\alpha_j| \le q^{1/2}$.

Let $L(t) = \prod_{j=1}^{2g} (1 - \alpha_j t)$ and consider the power series

$$\log \frac{1}{L(t)} = \sum_{d=1}^{\infty} (\alpha_1^d + \cdots + \alpha_{2g}^d) \frac{t^d}{d}.$$

By (ii) we have

$$\left| \log \frac{1}{L(t)} \right| \le \sum_{d=1}^{\infty} (A + Bq^{d/2}) \frac{|t|^d}{d}$$

$$\le A \log \frac{1}{(1 - |t|)} + B \log \frac{1}{(1 - |q^{1/2}t|)};$$

hence the power series converges absolutely for any complex number t that satisfies $|t| < q^{-1/2}$. Therefore all the singularities of $\log L(t)^{-1}$ lie in the region $|t| \ge q^{-1/2}$; in particular the zeros of the polynomial $L(t)$ satisfy

$$|1/\alpha_j| \ge q^{-1/2}$$

and equivalently $|\alpha_j| \le q^{1/2}$. This proves the lemma.

We now make some preparations for the next lemma which contains the basic idea from Galois theory which will be used in the proof of the RH.

Let C_0 be a non-singular complete connected curve rational over the field $\mathbb{F}_0 = \mathbb{F}_q$ and with function field K_0. Let $\mathbb{F}$ be the algebraic closure of $\mathbb{F}_0$ and let $K = K_0 \cdot \mathbb{F}$ be the compositum of K_0 and $\mathbb{F}$. Denote by

$$\varphi : K \to K$$

The Frobenius automorphism which acts like the identity on the constants and raises the non-constant elements to the q-th power: $\varphi(f) = f^q$. Let C be the smooth model corresponding to the function field K. K contains a purely transcendental subfield $\mathbb{F}(t)$ such that K is a separable extension of $\mathbb{F}(t)$. Hence there is a normal extension $\mathbb{F}(t)$ which is also normal over K. Geometrically this means that there is a smooth curve C' with function field K' which is a Galois covering of C and of $\mathbb{P}^1$:

$$C' \to C \to \mathbb{P}^1.$$

These coverings are unramified outside a finite set of points of the projective line $\mathbb{P}^1$. We let G be the group of automorphisms of $C'/\mathbb{P}^1$ and let H be the subgroup of those which act trivially on C. The group G may not act on C' over $\mathbb{F}_0$ but it always acts over a finite extension of $\mathbb{F}_0$, and therefore for our purposes we assume, as we may use Corollary 3.3.1, that G acts on C' over $\mathbb{F}_0$.

If P is a point of $\mathbb{P}^1$ rational over $\mathbb{F}$ and unramified in $C' \to \mathbb{P}^1$, and if Q is a point of C' lying over P, we have

$$\varphi(Q) = g \cdot Q$$

for some $g \in G$, called the *Frobenius substitution* of G at the point Q. Let $N_1(C', g)$ be the number of such points of C' with Frobenius substitution g.

Lemma 3.5 *With the notations above we have*

$$\sum_{g \in G} N_1(C', g) = |G| N_1(\mathbb{P}^1) + \mathcal{O}(1),$$

where the implied constant is independent of q and $N_1(\mathbb{P}^1)$ is the number of points P on $\mathbb{P}^1(\mathbb{F})$ with $\varphi(P) = P$.

Proof. Let P be a closed point in $\mathbb{P}^1$ unramified in $f\colon C' \to \mathbb{P}^1$ and denote by $f^{-1}(P) = \{Q_1, \ldots, Q_R\}$ the distinct points of C' which lie over P.

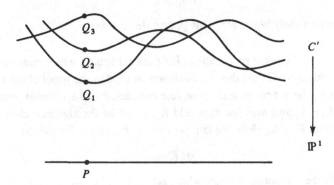

Figure 3.3

Let $Q \in f^{-1}(P)$ and suppose $\varphi(Q) = g \cdot Q$ for some $g \in G$. Since the Frobenius automorphism commutes with projection, that is $\varphi \circ f = f \circ \varphi$, it follows that $\varphi(P) = P$. This leads to the problem of determining the closed points P on $\mathbb{P}^1(\mathbb{F})$ with $\varphi(P) = P$, i.e. the fixed points of Frobenius. We claim that these are extensions of the closed points in $\mathbb{P}^1(\mathbb{F}_0)$ of degree 1 over $\mathbb{F}_0$. In fact the closed point P in $\mathbb{P}^1(\mathbb{F})$ corresponds to a valuation of $\mathbb{F}(t)$ with residue class field $\mathbb{F}_P = \mathbb{F}_{OP} \circ \mathbb{F}$. Now, since the Frobenius morphism φ acts like the identity on $\mathbb{F}$ and also $\varphi(P) = P$, we have that φ also acts like the identity on $\mathbb{F}_{OP}$. But $\mathbb{F}_{OP}$ is a finite extension of $\mathbb{F}_0$ and clearly φ acts like the identity on $\mathbb{F}_{OP}$ if and only if $[\mathbb{F}_{OP} : \mathbb{F}_0] = 1$, that is the projection of P to $\mathbb{P}^1(\mathbb{F}_0)$ is a closed point of degree 1 over $\mathbb{F}_0$.

Let $\{P_1, \ldots, P_s\}$ be the points of $\mathbb{P}^1(\mathbb{F})$ which lie above the closed points in $\mathbb{P}^1(\mathbb{F}_0)$ of degree 1 over $\mathbb{F}_0$. Let P be a point of $\mathbb{P}^1(\mathbb{F})$ which is unramified in $C' \xrightarrow{f} \mathbb{P}^1$ and let Q be a point of C' lying over P, and denote by

$$D(Q) = \{\sigma \in G \,|\, \sigma Q = Q\}$$

its decomposition group. By Galois theory we know G acts transitively on the fiber $f^{-1}(P)$ which is also the orbit of Q under the action of G. Therefore the number of elements in the orbit of Q is equal to the index of $D(Q)$ in G:

$$|G| = |D(Q)| |f^{-1}(P)|.$$

On the other hand, if $\varphi(Q) = g \cdot Q$, then $\varphi(Q) = gh \cdot Q$ for any $h \in D(Q)$. Therefore if Q is counted in $N_1(C', g)$, it is also counted in $N_1(C', gh)$ for any $h \in D(Q)$. It follows then that any of the points in the orbit of Q under the action of G is counted $|D(Q)|$ times among the numbers $N_1(C', g)$ as g ranges over G. Since each element in the orbit of Q is counted $|D(Q)|$ times in the sum

$$\sum_{g \in G} N_1(C', g),$$

the total contribution to the sum is $|D(Q)| |f^{-1}(P)| = |G|$. Now with at most a bounded number of exceptions, all the $N_1(\mathbb{P}^1)$ points of $\mathbb{P}^1(\mathbb{F})$ with $\varphi(P) = P$ contribute the same amount to the sum above. Thus we finally obtain

$$\sum_{g \in G} N_1(C', g) = |G| N_1(\mathbb{P}^1) + \mathcal{O}(1),$$

where the $\mathcal{O}(1)$ takes care of the branch points of $C' \to \mathbb{P}^1$ where the equality $|f^{-1}(P)| = [G : D(Q)]$ does not hold. This proves the lemma.

Remark A similar reasoning, replacing $\mathbb{P}^1$ by C, also proves

$$\sum_{h \in H} N_1(C', h) = |H| N_1(C) + \mathcal{O}(1). \tag{3.2}$$

Proof of the RH for a Galois covering. Suppose for the moment that $C \to \mathbb{P}^1$ is a Galois covering and assume that for a given automorphism g

of $C/\mathbb{P}^1$ there exist constants A and B independent of q such that

$$N_1(C, g) \leq q + 1 + A + Bq^{1/2}. \tag{3.3}$$

Then by isolating the term $N_1(C, 1)$ in the sum (3.2), which corresponds to the identity element in G, and substituting for the other elements the estimate (3.3) we get

$$N_1(C, 1) \geq |G|(1 + q) - (|G| - 1)(1 + q + A + Bq^{1/2}) + \mathcal{O}(1);$$

equivalently we get

$$N_1(C, 1) \geq q + 1 + A' + B'q^{1/2}. \tag{3.4}$$

The estimates (3.3) with $g = $ identity and (3.4) give

$$|N_1(C, 1) - (q + 1)| \leq A'' + B''q^{1/2}.$$

From this we conclude that if we knew that $C \to \mathbb{P}^1$ is a Galois covering and if we knew that the estimates (3.3) hold for all $g \in \mathrm{Aut}(C/\mathbb{P}^1)$, then the RH for the curve $C/\mathbb{F}_q$ would follow.

The proof of (3.3) which we give below is entirely due to Bombieri and rests on an idea of Stepanof.

Lemma 3.6 (Bombieri) *Let $q = p^\alpha$ with α even satisfy $q > (g + 1)^4$. Assume $C \to \mathbb{P}^1$ is a Galois covering defined over $\mathbb{F}_q$. Let $g \in \mathrm{Aut}(C/\mathbb{P}^1)$ and put*

$$N_1(C, g) = \{Q \in C(\mathbb{F}): \varphi(Q) = g \cdot Q\}.$$

Then we have

$$N_1(C, g) \leq 1 + q + (2g + 1)q^{1/2}.$$

Proof. If $N_1(C, g) = 0$, there is nothing to prove; hence we may assume that C has a point P that satisfies

$$\varphi(P) = g \cdot P.$$

In the following we keep P fixed. Let m be a positive integer and put

$$L_m = \{f \in K | (f) + mP \geq 0\} = L(mP).$$

As in Chapter 2, L_m is considered here as a vector space over $\mathbb{F}$ by adding the function which is identically zero. A function different from 0 belongs to the space L_m if it has at most a pole at P of order not exceeding m. In the following we put

$$l_m = \dim_\mathbb{F} L_m.$$

We observe that the Riemann–Roch theorem implies that

$$l_m \geq m + 1 - g \quad \text{and} \quad l_m = m + 1 - g \quad \text{if } m > 2g - 2.$$

By considering the expansion of a function in terms of a local uniformizing parameter at the point P we see easily that $l_m \leq m + 1$. Put $\Phi = g^{-1} \cdot \varphi$ and define a new space of functions by

$$L_m^{(\Phi)} = \{f \circ \Phi \mid f \in L_m\}.$$

If we recall that the automorphism $\Phi = g^{-1} \cdot \varphi$ acts like the identity on the field of constants $\mathbb{F}$, we obtain a sequence of maps

$$L_m \overset{\Phi}{\underset{\approx}{\to}} L_m^{(\Phi)} \to L_{mq},$$

where the last inclusion results by observing that every element of $L_m^{(\Phi)}$ is a q-th power; in fact at the level of divisors we have $(f \circ \Phi) = q\Phi((f))$.

In the following we need to consider the absolute Frobenius map

$$F_{\mathrm{abs}} \colon K \to K$$

which acts on the functions by raising them to the p-th power. Now for an integer $l \geq 1$ we define

$$L_l^{(p^\mu)} = \{f \circ (F_{\mathrm{abs}})^\mu \mid f \in L_l\}$$

and observe that

$$L_l \overset{\sim}{\to} L_l^{(p^\mu)} \to L_{lp^\mu}.$$

In the proof of Lemma 3.6 we need the following subsidiary.

Lemma 3.7 *If $lp^\mu < q$, then the multiplication map*

$$L_l^{(p^\mu)} \otimes_{\mathbb{F}} L_m^{(\Phi)} \to L_{lp^\mu + mq}$$

is injective.

Proof. We first observe that the space L_m has a basis $f_1, \ldots, f_r$ satisfying

$$\mathrm{ord}_P f_i < \mathrm{ord}_P f_{i+1} \quad \text{for } i = 1, \ldots, r - 1.$$

In fact, since we have a filtration $(0) \subset \mathbb{F} = L_0 \subset L_1 \subset \cdots \subset L_m$, it follows that

$$L_m \simeq \bigoplus_{i=0}^{m} L_i / L_{i-1};$$

from the inequality $\dim_{\mathbb{F}} L_{m+1} \leq \dim_{\mathbb{F}} L_m + 1$ it also follows that $\dim L_i / L_{i-1} \leq 1$; the basis $f_1, \ldots, f_r$ is then obtained by picking up for each i, when possible, one element of L_i not in L_{i-1}.

Now the lemma claims that a function of the form

$$F(x) = s_1^{p^\mu}(x) f_1^{(\Phi)}(x) + \cdots + s_r^{p^\mu}(x) f_r^{(\Phi)}(x) \qquad \text{with } s_1, \ldots, s_r \in L_l,$$

is identically zero only if all the $s_i(x)$ are identically 0. In fact suppose that $F(x)$ is identically 0 and that $s_h(x)$ is the first s_i which is not identically zero. Taking the order at P of both sides of the identity

$$s_h^{p^\mu}(x) f_h^{(\Phi)}(x) = -s_{h+1}^{p^\mu}(x) f_{h+1}^{(\Phi)}(x) - \cdots - s_r^{p^\mu}(x) f_r^{(\Phi)}(x)$$

we obtain using the properties of the basis elements $f_1, \ldots, f_r$ that

$$p^\mu \operatorname{ord}_P s_h + q \operatorname{ord}_P f_h \geq \min_{i > h} (p^\mu \operatorname{ord}_P s_i + q \operatorname{ord}_P f_i)$$

$$\geq -p^\mu l + q \operatorname{ord}_P f_{h+1},$$

therefore

$$p^\mu \operatorname{ord}_P s_h \geq -l p^\mu + q(\operatorname{ord}_P f_{h+1} - \operatorname{ord}_P f_h)$$

$$\geq -l p^\mu + q > 0.$$

This means that s_h vanishes at P, and thus is a function with no poles and at least one zero, hence s_h is identically 0, contradicting our original assumption. This proves Lemma 3.7.

We now define a new map δ by means of the following diagram:

$$
\begin{array}{ccccc}
L_l^{(p^\mu)} \cdot L_m^{(\Phi)} & \xrightarrow{\;\delta\;} & L_l^{(p^\mu)} \cdot L_m & \hookrightarrow & L_{lp^\mu + m} \\
\wr \downarrow & & \uparrow {\scriptstyle \text{multiplication}} & & \\
L_l^{(p^\mu)} \otimes_{\mathbb{F}} L_m^{(\Phi)} & \xleftarrow{\;\sim\;} & L_l^{(p^\mu)} \otimes_{\mathbb{F}} L_m. & &
\end{array}
$$

Let us suppose that we can arrange to have the following inequalities.

(i) $l p^\mu < q$
(ii) $l_m l_l > l_{m + l p^\mu}$
(iii) $l + mq/p^\mu + 1 \leq 1 + q + (2g + 1)q^{1/2}$.

From (i) and Lemma 3.7 we see that the map δ is well defined; (ii) implies that the kernel of δ is not empty and so must contain a non-zero function F satisfying

$$F(x) = \sum_{i=1}^{r} s_i^{p^\mu}(x) f_i^{(\Phi)}(x) \qquad \text{but} \qquad \delta(F) = \sum_{i=1}^{r} s_i^{p^\mu}(x) f_i(x) = 0.$$

Also we observe that since $p^\mu < q$ and $f_i^{(\Phi)}$ is a q-th power, the function F itself is a p^μ-th power. Therefore if the point Q is different from P and satisfies

$$\varphi(Q) = g \cdot Q, \qquad \text{or} \qquad \Phi(Q) = Q,$$

then

$$\mathrm{ord}_Q F = \mathrm{ord}_Q\left(\sum_i s_i^{p^\mu}(x) f_i^{(\Phi)}(x)\right) = \mathrm{ord}_Q\left(\sum_i s_i^{p^\mu}(x) f_i(x)\right) > 0;$$

hence we have

$$p^\mu(N_1(C,g) - 1) \le \deg(F)_0 = \deg(F)_\infty \le lp^\mu + mq,$$

where $(F)_0$ denotes the divisor of zeros of F and $(F)_\infty$ the divisor of poles. From the inequality (iii) we obtain

$$N_1(C;g) \le l + mq/p^\mu + 1 \le 1 + q + (2g + 1)q^{1/2}.$$

To complete the proof of Lemma 3.6 it remains to verify the inequalities (i), (ii) and (iii); we now do this in the following

Lemma 3.8 *If we take* $\mu = \alpha/2$, *so that* $p^\mu = q^{1/2}$, $m = q^{1/2} + 2g$ *and* $l = g + 1 + [g/(g + 1)q^{1/2}]$, *then the inequalities*

(i) $lp^\mu < q$,
(ii) $l_m l_l > l_m + lp^\mu$ *and*
(iii) $l + mq/p^\mu + 1 \le 1 + q + (2g + 1)q^{1/2}$

are satisfied.

Proof. To verify (iii) we use the assumption $q > (g + 1)^4$ in the form

$$q^{1/2}/(g + 1) - g - 1 > 0$$

to obtain

$$\begin{aligned}
l + mqp^{-\mu} + 1 &= g + 1 + [q^{1/2}g/(g + 1)] + (q^{1/2} + 2g)q^{1/2} + 1 \\
&\le g + 1 + q^{1/2}g/(g + 1) + q + 2gq^{1/2} + 1 \\
&\le 1 + q + (2g + 1)q^{1/2} - (q^{1/2}/(g + 1) - g - 1) \\
&\le 1 + q + (2g + 1)q^{1/2}.
\end{aligned}$$

To verify (i) we observe also that

$$\begin{aligned}
l &= [q^{1/2}g/(g + 1)] + g + 1 + q^{1/2}/(g + 1) - q^{1/2}/(g + 1) \\
&\le p^\mu - (q^{1/2}/(g + 1) - g - 1) \\
&\le p^\mu.
\end{aligned}$$

To verify (ii) we observe that since $l, m \ge 2g$, the Riemann–Roch theorem gives equalities for the dimensions l_l, l_m and l_{m+lp^μ} and so it suffices to show that

$$(l + 1 - g)(m + 1 - g) > l_{m+lp^\mu} = m + lp^\mu + 1 - g.$$

To verify this we consider the equivalent inequalities

$$(l - g)(m + 1 - g) > lp^\mu,$$

$$l(m + 1 - g - p^\mu) > g(m + 1 - g),$$

$$l(g + 1) > g(q^{1/2} + g + 1),$$

$$l > q^{1/2}g/(g + 1) + g,$$

and this last inequality follows from the definition of l. This completes the proof of Lemma 3.6.

Reduction to the case of a Galois covering

Let $C_0/\mathbb{F}_q$ be a curve with function field K_0. Let $K = K_0 \cdot \mathbb{F}$. Let t be a non-constant element in K so that K_0 is a separable extension of $\mathbb{F}_q(t)$. Let K' be the smallest normal extension of $\mathbb{F}_q(t)$ which contains K and let $G = \text{Gal}(K'/\mathbb{F}_q(t))$. If necessary we may replace $\mathbb{F}_q$ by $\mathbb{F}_{q^N}$ and assume K' has the exact field of constants $\mathbb{F}_q$. Now K is the fixed field of some subgroup H of G and K' is a Galois extension of K with Galois group H. Extending the field of constants from $\mathbb{F}_q$ to its algebraic closure $\mathbb{F}$ and denoting by C' a smooth model with function field K', we have Galois coverings

$$C' \to C \to \mathbb{P}^1.$$

In particular by the remark following the proof of Lemma 3.5 we obtain

$$N_1(C) = |H|^{-1} \sum_{h \in H} N_1(C', h) + \mathcal{O}(1),$$

where $N_1(C)$ denotes the number of closed points P in C which satisfy $\varphi(P) = P$, i.e. the points fixed by Frobenius. This equality shows that the Riemann Hypothesis for the curve $C_0/\mathbb{F}_q$ will follow if it can be shown that for any $h \in H$

$$|N_1(C', h) - (q + 1)| \ll q^{1/2}, \tag{3.5}$$

this is clear if we substitute the above estimate into the representation of $N_1(C)$ to obtain

$$|N_1(C) - (q + 1)| \leq A + Bq^{1/2}.$$

To obtain (3.5) we recall that Lemma 3.6 gives

$$N_1(C', h) \leq q + 1 + A + Bq^{1/2}.$$

Now Lemma 3.5 applied to the covering $C' \to \mathbb{P}^1$ gives

$$(q + 1)|G| = \sum_{g \in G} N_1(C', g) + \mathcal{O}(1)$$

and this can hold if and only if

$$N_1(C', h) = q + \mathcal{O}(q^{1/2}).$$

This completes the proof of the Riemann hypothesis for the curve $C/\mathbb{F}_q$.

3.5 The *L*-functions of curves and their functional equations

3.5.1 Preliminary remarks and notation

In this section we present the proof of the functional equation for the *L*-functions of algebraic curves due to Schmid and Teichmuller [68] (for a precise statement see Theorem 3.4 below). In an attempt to convey the clarity of Teichmuller's proof we will divide the discussion in two parts; in the first we shall collect those aspects of the argument which are of a purely algebraic or combinatorial nature; the second will be of a geometric nature and relies heavily on such results as the Riemann–Roch theorem and the residue theorem.

It is perhaps appropriate that we say here a few words about the relevance of the *L*-functions for number theoretic purposes as well as the significance of the methods used. Most of the *L*-functions studied in number theory contain in their definition two aspects which are more or less well understood, i.e. the additive versus the multiplicative definition as exemplified in the equality between a Dirichlet series and an Euler product. In our situation this duality takes the form

$$\sum_D \chi(D)N(D)^{-s} = \prod_P (1 - \chi(P)N(P)^{-s})^{-1}.$$

A third and deeper aspect of the *L*-functions is the analogue of the Hadamard product formula for the Riemann zeta function, which in the case of the *L*-functions of algebraic curves takes the form of a representation as a polynomial; more precisely as the characteristic polynomial of a Frobenius substitution. This property of the function $L(s, \chi)$, which depends on the construction of certain *l*-adic vector spaces associated with the curve C, will not be persued here. We do prove that $L(s, \chi)$ is a polynomial in q^{-s} with a degree that can be effectively calculated from C and χ. To a first approximation we have here a hint of the existence of a relation such as the Lefschetz–Weil–Grothendieck fixed-point formula, which is the principal tool in the more advanced study of the *L*-functions.

From the point of view of applications to number theory and coding theory, the study of certain Galois coverings of the curve C, together with the bounds already obtained for the zeros of the zeta functions of such coverings, leads to sharp estimates for a wide class of exponential sums, e.g. the Kloosterman sums.

The adelic proof of the functional equation, following the techniques of Tate, which one sees more frequently nowadays, did not seem to us to be appropriate to our elementary treatment of the subject of algebraic curves. For these reasons we decided in favor of the proof of Schmid and Teichmuller. Nevertheless, we have indicated in several remarks how the arguments and results can be reformulated in terms of the elementary language of harmonic analysis on finite abelian groups. Another aspect of the proof which is worth mentioning is the appearance of a finite commutative ring R which also plays an important role in the study of the generalized Jacobians of the curve C. We do not go here into a closer study of this analogy, which we expect contains information about the ramified coverings of the curve C.

Notation

Let $k = \mathbb{F}_q = \mathbb{F}$ be a finite field with q elements. Let C be a nonsingular, proper, and geometrically connected curve defined over k. Let K be its function field and k its exact field of constants. Let S be a finite set of indices and let $\{P_v\}_{v \in S}$ be a finite set of distinct closed points on the curve C; let $\{e_v\}_{v \in S}$ be a finite set of positive integers. With the above data we associate a positive divisor on the curve C:

$$F = \sum_{v \in S} e_v P_v.$$

For each $v \in S$, let k_v denote the residue field of P_v and let $f_v = \deg(P_v)$, i.e. $f_v = [k_v : k]$. We clearly have $\deg F = \sum_{v \in S} e_v f_v$; in the following we denote this degree by f.

3.5.2 Algebraic aspects

For each $v \in S$ consider a commutative ring of the form

$$R_v = k_v + k_v t_v + \cdots + k_v t_v^{e_v - 1}, \qquad \text{with } t_v^{e_v} = 0.$$

Denote the direct sum of the rings R_v by $R = \bigoplus_{v \in S} R_v$. An arbitrary element r in R can be expressed in the form

$$r = \sum_{v \in S} r_v, \qquad \text{with } r_v = \sum_i a_{v,i} t_v^i, \tag{3.6}$$

where $0 \le i < e_v$ and the $a_{v,i}$ are elements in k_v. The ring R considered as a vector space over k has dimension

$$\dim_k R = \deg F = f.$$

For an element r in R we define the trace function $\mathrm{Tr}(r)$ by

$$\mathrm{Tr}(r) = \sum_{v \in S} \mathrm{Tr}(r_v), \qquad \text{and} \qquad \mathrm{Tr}(r_v) = \mathrm{Tr}_{k_v/k}(a_{v,e_v-1}),$$

where r_v is the v-th component of r and $a_{v,e_v-1} \in k_v$ is the $(e_v - 1)$-th coefficient of r_v, $v \in S$, which appears in the decomposition in (3.6); $\mathrm{Tr}_{k_v/k}$ denotes the ordinary trace from k_v to k.

It is easily verified that for any elements r and s in R and a in $\mathbb{F}_q$ we have

$$\mathrm{Tr}(r + s) = \mathrm{Tr}(r) + \mathrm{Tr}(s), \qquad \mathrm{Tr}(ar) = a\,\mathrm{Tr}(r).$$

Lemma 3.9 *The form $\langle x, y \rangle = \mathrm{Tr}(xy)$ defines a non-degenerate $\mathbb{F}$-bilinear pairing*

$$\langle \ , \ \rangle \colon R \times R \to \mathbb{F}_q,$$

i.e. *if $x \in R$ is such that $\langle x, y \rangle = 0$ for all $y \in R$, then $x = 0$.*

Proof. Suppose we have $x \ne 0$. Then in the representation (3.6), we have a coefficient a_j different from 0 with a minimal j. If we put $y = a_v a_{v_j}^{-1} t_v^{e_v-1-j}$, where a_v is an element in $\mathbb{F}_v$ with $\mathrm{Tr}_{\mathbb{F}_v/\mathbb{F}}(a_v) \ne 0$, then $xy = a_v t_v^{e_v-1}$ and we obtain

$$\langle x, y \rangle = \mathrm{Tr}(xy) = \mathrm{Tr}_{\mathbb{F}_v/\mathbb{F}}(a_0) \ne 0,$$

which contradicts the assumption that $\langle x, y \rangle = 0$ for all y.

Let M be an $\mathbb{F}$-submodule in R. With M we associate another $\mathbb{F}$-submodule $M^\perp$ which we call the complementary module.

Definition 3.2 $M^\perp = \{y \in R \mid \langle x, y \rangle = 0 \text{ for all } x \in M\}.$

Lemma 3.10 *We have*

$$\dim_{\mathbb{F}} M + \dim_{\mathbb{F}} M^\perp = \dim_{\mathbb{F}} R = f.$$

Proof. Let $x_1, \ldots, x_m$ be an $\mathbb{F}$-basis for M and $y_1, \ldots, y_f$ a similar basis for $M^\perp$. Any element y in $M^\perp$ can be expressed in the form

$$y = b_1 y_1 + \cdots + b_f y_f, \qquad b_1, \ldots, b_f \in \mathbb{F}.$$

The conditions $\langle x, y \rangle = 0$ are equivalent to a homogeneous system of linear equations

$$b_1\langle x_\mu, y_1 \rangle + \cdots + b_f\langle x_\mu, y_f \rangle = 0, \qquad 1 \le \mu \le m, \qquad (3.7)$$

for the coefficients $b_1, \ldots, b_f$. By Lemma 3.9 we know that the rank of the matrix $(\langle x_\mu, y_\nu \rangle)_{\substack{1 \le \mu \le m \\ 1 \le \nu \le f}}$ is equal to m. Therefore the dimension of the solution space of the system in (3.7) is $f - m$; i.e. $\dim_\mathbb{F} M + \dim_\mathbb{F} M^\perp = f$. Let $M^{\perp\perp}$ be the complementary module of $M^\perp$. Then clearly $M \subset M^{\perp\perp}$; the previous dimension relation implies $M = M^{\perp\perp}$.

For each v in S, the zero divisors in the ring R_v are those elements of the form $a_1 t_v + \cdots + a_{e_v - 1} t_v^{e_v - 1}$ with $a_i \in \mathbb{F}_v$. We also observe that an element $x = \sum_v x_v$ in R is a zero divisor if and only if at least one of the components x_v is a zero divisor in R_v.

Let χ be a character of the multiplicative group of all non-zero divisors in R. We extend the definition of χ to all of R by putting $\chi(x) = 0$ if x is a zero divisor. For the applications we have in mind the characters of interest are those which satisfy

$$\chi(a) = 1 \qquad \text{for } a \in \mathbb{F}. \qquad (3.8)$$

Definition 3.3 *With each $\mathbb{F}$-submodule M in R and each character χ as defined above, we put*

$$\Lambda_\chi(M) = \sum_{x \in M} \chi(x). \qquad (3.9)$$

In the following we consider a function $e: \mathbb{F} \to \mathbb{C}$ subject to the conditions

$$e(0) = 1 \quad \text{and} \quad \sum_{a \in \mathbb{F}} e(a) = 0. \qquad (3.10)$$

We now make the following assumptions on the character χ. For each v in S there is an element $a_v \in \mathbb{F}_v$ with

$$\chi(1 + a_v t_v^{e_v - 1}) \ne 1. \qquad (3.11)$$

In case $e_v = 1$ we only suppose that $1 + a_v$ is not a zero divisor. We also assume that in each case

$$\chi(1 + a_v t_v^{e_v - 1}) \ne 0. \qquad (3.12)$$

Remark In the geometric situation to be considered in Section 3.5.3, the above assumptions amount to the fact that χ is a primitive character modulo F, i.e. the divisor $F = \sum_{v \in S} e_v P_v$ is the conductor of χ.

Lemma 3.11 *For any* $\mathbb{F}$-*submodule* M *of* R *we have*

$$\sum_{x \in M} e(\mathrm{Tr}(xy)) = \begin{cases} q^{\dim_\mathbb{F} M} & \text{if } y \in M^\perp \\ 0 & \text{otherwise.} \end{cases} \tag{3.13}$$

Proof. 1. By definition $y \in M^\perp$ implies that $\mathrm{Tr}(xy) = 0$ for all $x \in M$ and hence

$$\sum_{x \in M} e(\mathrm{Tr}(xy)) = \mathrm{Card}\, M = q^{\dim_\mathbb{F} M}.$$

2. Let $x_1, \ldots, x_m$ be an $\mathbb{F}$-basis of M. If y does not belong to $M^\perp$, then for at least one index μ $(1 \le \mu \le m)$ we have $\mathrm{Tr}(x_\mu y) \ne 0$. Without loss of generality we may suppose that $\mathrm{Tr}(xy) \ne 0$. We can now write

$$\sum_{x \in M} e(\mathrm{Tr}(xy)) = \sum_{(a_i)} e\left(\mathrm{Tr}\left(\sum_\mu a_\mu x_\mu \cdot y\right)\right)$$

$$= \sum_{(a_i)} e\left(\sum_\mu a_\mu \mathrm{Tr}(x_\mu \cdot y)\right),$$

where (a_i) runs over all m-tuples of elements $(a_1, \ldots, a_m)$ in $\mathbb{F}^m$. Now, as a_1 ranges over $\mathbb{F}$, for a fixed $(m-1)$-tuple $a_2, \ldots, a_m$, the expression $\sum_\mu a_\mu \mathrm{Tr}(x_\mu \cdot y)$ also runs over all the elements of $\mathbb{F}$ exactly once; hence by (3.10) the entire sum is 0.

Definition 3.4 *To a character* χ *and a function* e *we associate a Gauss sum*

$$G(x) = \sum_{y \in R} \bar\chi(y) e(\mathrm{Tr}(xy)).$$

Lemma 3.12 *For all* x *in* R *we have*

$$G(x) = \chi(x) G(1). \tag{3.14}$$

Proof. (i) We suppose x is not a zero divisor in R and make the substitution $xy = z$; as y runs over all the elements in R, and x is kept fixed, z also runs over all elements in R and conversely. Thus we have

$$G(x) = \sum_{z \in R} \bar\chi(x^{-1}z) e(\mathrm{Tr}(x)) = \chi(x) \sum_{z \in R} \bar\chi(z) e(\mathrm{Tr}(z))$$

$$= \chi(x) G(1).$$

(ii) If x is a zero divisor in R, then at least one of the components in the decomposition

$$x = \sum_v x_v,$$

say x_v, is a zero divisor. The assumptions (3.11) and (3.12) imply that there is a non-zero divisor

$$z_v = 1 + a_v t_v^{e_v - 1} \qquad (a_v \in \mathbb{F}_v)$$

with $\chi(z_v) \neq 1$. For these two elements we clearly have $xz_v = x$ and

$$\chi(z_v)G(x) = \sum_{y \in R} \bar{\chi}(z_v^{-1}y)e(\mathrm{Tr}(xy)).$$

If we make the change of variable $y' = z_v^{-1}y$, then we obtain

$$\chi(z_v)G(x) = \sum_{y' \in R} \bar{\chi}(y')e(\mathrm{Tr}(xz_v y'))$$

$$= \sum_{y' \in R} \bar{\chi}(y')e(\mathrm{Tr}(xy'))$$

$$= G(x).$$

Therefore we have $G(x) = 0$ and (3.14) holds in all cases.

If M is a subset of R we put

$$\|M\| = \mathrm{Card}\, M.$$

In particular, when M is an $\mathbb{F}$-submodule we have $\|M\| = q^{\dim_\mathbb{F} M}$.

Lemma 3.13 *There is a function $W(\chi)$ which depends only on R and the character χ such that for every $\mathbb{F}$-submodule M in R we have*

$$\|M\|\Lambda_{\bar{\chi}}(M) = W(\chi)\Lambda_\chi(M).$$

(Here $\bar{\chi}$ is the complex conjugate of χ.)

Corollary 3.13.1 *In the notation of Lemma 3.12 we have*

$$W(\chi) = G(1) = \sum_{z \in R} \bar{\chi}(z)e(\mathrm{Tr}(z)).$$

Proof of Lemma 3.13. We will obtain the desired equality by evaluating the double sum

$$S = \sum_{x \in M} \sum_{y \in R} \bar{\chi}(y)e(\mathrm{Tr}(xy))$$

in two different ways. First by inverting the order of summation and using Lemma 3.11 we obtain

$$S = \sum_{y \in R} \bar{\chi}(y) \sum_{x \in M} e(\mathrm{Tr}(xy))$$

$$= \sum_{y \in M} \chi^{-1}(y)\|M\|$$

$$= \|M\|\Lambda_{\chi^{-1}}(M).$$

Secondly from the definition of the Gauss sum we have

$$S = \sum_{x \in M} G(x)$$

$$= \sum_{x \in M} \chi(x) W(\chi)$$

$$= W(\chi) \Lambda_\chi(M).$$

The proof of the lemma is now complete.

Remarks 1. If in Lemma 3.13 we exchange the role of χ and $\bar{\chi}$ and of M and $M^\perp$ we then obtain

$$\|M^\perp\| \Lambda_\chi(M) = W(\bar{\chi}) \Lambda_{\chi^{-1}}(M^\perp).$$

The assumption $\chi(a) = 1$ for all $a \in \mathbb{F}^\times$ gives that for the trivial module $M = \mathbb{F}$, $\Lambda_\chi(M) = q - 1$. Hence $\Lambda_\chi(M)$ is not identically zero. We can then combine the last equality with that of Lemma 3.13 and obtain

$$\|M\| \Lambda_\chi(M) = W(\bar{\chi}) W(\chi) \Lambda_\chi(M) \|M\|^{-1},$$

from which it follows by Lemma 3.10 that

$$W(\chi^{-1}) W(\chi) = \|M\| \|M^\perp\| = q^f. \tag{3.15}$$

In the special case $M = k$, we have $M = \{y \in R | \text{Tr}(y) = 0\}$ and also

$$\Lambda_\chi(M) = q - 1 \qquad \text{and} \qquad \Lambda_{\bar{\chi}}(M) = \sum_{\substack{y \in R \\ \text{Tr}(y)=0}} \bar{\chi}(y).$$

Lemma 3.13 applied in this case gives

$$q \sum_{\substack{y \in R \\ \text{Tr}(y)=0}} \bar{\chi}(y) = W(\chi)(q - 1).$$

Thus $W(\chi)$ has the representation

$$W(\chi) = \frac{q}{q - 1} \sum_{\substack{y \in R \\ \text{Tr}(y)=0}} \bar{\chi}(y);$$

this implies that

$$\overline{W}(\chi) = W(\bar{\chi}). \tag{3.16}$$

Combining (3.15) and (3.16) we obtain

$$|W(\chi)|^2 = q^f.$$

2. If we introduce a measure μ on the ring R, we can then interpret the various results obtained above from the point of view of the elementary

harmonic analysis on the finite abelian group R. Let us first consider the space of functions

$$S(R) = \left\{ \sum_M c_M f_M \,\middle|\, M \text{ an } \mathbb{F}\text{-submodule of } R, \, c_M \in \mathbb{C} \right.$$

$$\text{and } f_M \text{ the characteristic function of } M \bigg\}.$$

Let ψ be an additive character of R chosen so that the bilinear pairing

$$\langle\ ,\ \rangle \colon R \times R \to \mathbb{C}$$

$$\langle x, y \rangle = \psi(xy)$$

is non-degenerate. We define the Fourier transform of a function $f \in S(R)$ by

$$\hat{f}(y) = \int_R \psi(xy) f(x) \, d\mu.$$

If we observe that for any $\mathbb{F}$-submodule M we have

$$\hat{f}_M = \|M\| f_{M^\perp}$$

and that

$$\Lambda_\chi(M) = \int_R \chi(x) f_M(x) \, d\mu,$$

Then the statement of Lemma 3.13 can be phrased as saying that there is a constant $W(\chi)$ which depends only on χ and R such that for any $f \in S(R)$ we have

$$\int_R \chi^{-1}(a) \hat{f}(a) \, d\mu = W(\chi) \int_R \chi(a) f(a) \, d\mu.$$

3.5.3 Geometric aspects

In the following C is a non-singular complete connected curve defined over the finite field of constants $\mathbb{F}$ with q elements. Let g be the genus of C and let K be its function field. Consider a primitive character of conductor F

$$\chi \colon \mathrm{Div}(C) \to \mathbb{C}^\times.$$

More precisely, let F denote the divisor ($\neq 0$)

$$F = \sum_{v \in S} e_v P_v,$$

where the P_v ($v \in S$) are closed points in C each of degree f_v, i.e. if $\mathbb{F}_v$ is the residue class field of P_v, then $\mathbb{F}_v$ is of degree f_v over $\mathbb{F}$. The primitivity of χ then means that

$$\text{if } \xi \in K^\times \text{ and } \operatorname{ord}_v(\xi - 1) \geq e_v \text{ for all } v \in S, \text{ then } \chi(\xi) = 1. \quad (3.17)$$

For each $w \in S$, there is an element $\xi_w \in K^\times$ whose divisor has support disjoint from S such that

$$\operatorname{ord}_v(\xi - 1) \geq e_v \text{ for all } v \neq w, \quad \text{but} \quad \operatorname{ord}_w(\xi - 1) = e_w - 1$$
$$\text{and} \quad \chi(\xi) \neq 1. \tag{3.18}$$

Let A_v be the local ring of P_v and M_v its maximal ideal. Let T_v be a local uniformizing parameter at P_v so that $M_v = T_v A_v$. For each v let t_v be the residue class of T_v under the reduction map

$$A_v \to A_v / M_v^{e_v}.$$

The intersection $A_S = \bigcap_{v \in S} A_v$ is the set of all functions in K which are integral at the points in S. The subring

$$A_S(F) = \{ f \in A_S | \operatorname{ord}_v f \geq e_v, \text{ for all } v \in S \}$$

is an ideal in A_S and the quotient ring is isomorphic to the direct sum

$$R = \bigoplus_{v \in S} R_v,$$

where each R_v is isomorphic to the residue class ring $R_v = A_v / M_v^{e_v}$; this last ring is of the form

$$R_v = \mathbb{F}_v + \mathbb{F}_v t_v + \cdots + \mathbb{F}_v t_v^{e_v - 1} \qquad \text{with } t_v^{e_v} = 0.$$

Thus the ring R has the structure of the rings studied in Section 3.5.2 and we can apply to R the theory developed there.

For the most part in the following we shall denote the elements of the function field K by Greek letters α, β, ... and their corresponding residue classes by small Latin letters a, b, ... etc.

We begin by observing that the canonical class contains a differential Ω, which at each closed point P_v, $v \in S$, has the principal part

$$T_v^{-e_v} dT_v.$$

This is in essence a Mittag–Leffler problem. To establish the existence of Ω we proceed as follows. Take an arbitrary differential Ω' whose associated divisor (Ω') is positive and split it in the form

$$(\Omega') = \sum_{v \in S} s_v P_v + N, \qquad S \cap \operatorname{Supp} N = \varnothing.$$

Now, let D be a positive divisor with support disjoint from S and with $\deg D$ so large that $\dim_F L((\Omega') - D) = 0$; then the Riemann–Roch theorem states that

$$\dim_F L(D) = \deg D + 1 - g.$$

If we take a point P_v, $v \in S$, and a positive integer s, then

$$\dim_F L(D + sP_v) - \dim_F L(D + (s-1)P_v) = \deg P_v.$$

This implies that there is a function $f_v^{(s)}$ with a pole of order exactly s at P_v and integral at the other points $Q \in S - \{P_v\}$. Furthermore, since the total number of such functions is equal to the number of elements in F_v, we can select $f_v^{(s)}$ with a Laurent expansion in powers of T_v which begins with a preassigned leading coefficient $a_v \in F_v$, i.e.

$$f_v^{(s)} = a_v T_v^{-s} + \cdots.$$

Now for each $v \in S$, select a function $f_v^{(e_v + s_v)}$ with the properties indicated and consider the differential

$$\Omega = \sum_{v \in S} f_v^{(e_v + s_v)}\Omega'.$$

If for each $v \in S$ we choose the leading coefficient of $f_v^{(e_v + s_v)}$ to be the inverse of the leading coefficient of Ω' we obtain that the principal part of Ω at v has the desired form $T_v^{-e_v} dT_v$. In the following we select Ω as above and keep it fixed; we also denote its divisor by

$$(\Omega) = W.$$

Lemma 3.14 *If* $\mathrm{Tr}\colon R \to F$ *is the* F*-linear function defined in* (3.5.2), *then we have for every* $\alpha \in A_S$

$$\mathrm{Tr}(a) = \sum_{v \in S} \mathrm{Res}_v(\alpha\Omega),$$

where a *is the residue class of* α *in* R.

Proof. If for each $v \in S$ we let

$$\alpha \equiv \sum_{i=0}^{e_v-1} a_{v,i} T_v^i \pmod{T_v^{e_v}}, \tag{3.19}$$

Then the residue class of α modulo $A_S(F)$ can be written in the form

$$a = \sum_{v \in S} \sum_{i=0}^{e_v-1} a_{v,i} T_v^i \tag{3.20}$$

and we have

$$\sum_{v \in S} \text{Res}_v(\alpha\Omega) = \sum_{v \in S} \text{Res}_v \left[\sum_{i=0}^{e_v-1} a_{v,i} T_v^i \frac{dT_v}{T_v^{e_v}} \right]$$

$$= \sum_{v \in S} \text{Tr}_{F_v/F}(a_{v,e_v-1})$$

$$= \text{Tr}(a),$$

which proves the lemma.

If $D \in \text{Div}(C)$ is an integral divisor we recall that the norm of D is defined to be $ND = q^{\deg D}$.

Definition 3.5 *The L-function associated with the character χ is defined by the series*

$$L(s, \chi) = \sum_D \chi(D) ND^{-s},$$

where the sum runs over all positive divisors $D \in \text{Div}(C)$ whose support is disjoint from F.

The Dirichlet series $L(s, \chi)$ can also be written in the form

$$L(s, \chi) = \sum_D \Theta_\chi(D) ND^{-s},$$

where the sum now runs over a representative system D of the absolute divisor classes which are prime to F, i.e. for a divisor D whose support is prime to F we have put

$$\Theta_\chi(D) = \sum_{D'} \chi(D'),$$

and the sum runs over all positive divisors $D' = D + (f)$ which are prime to F. Via the inclusion $K^\times \hookrightarrow \text{Div}(C)$ and the supposition $\chi(a) = 1$ for all $a \in F$, we can lift the character χ from the divisor group to functions; we can thus write

$$\Theta_\chi(D) = \frac{\chi(D)}{q - 1} \sum_\xi \chi(\xi),$$

where the sum runs over all elements in $K^\times$ whose divisors are disjoint from F and satisfy $(\xi) + D \geq 0$. If we agree to put $\chi(\xi) = 0$ for an element ξ with $S \cap \sup(\xi) \neq \varnothing$, then the last sum can also be written in the form

$$\Theta_\chi(D) = \frac{\chi(D)}{q - 1} \sum_{\xi \in L(D)} \chi(\xi).$$

Observe that since D is prime to F, $L(D) \subset A_S$. Now if two elements ξ, ξ' in $L(D)$ differ by an element in $L(D - F)$, then $\chi(\xi) = \chi(\xi')$. In fact, if

$$\xi - \xi' \in L(D - F),$$

then

$$\mathrm{ord}_v(\xi' - \xi) = \mathrm{ord}_v\left(\frac{\xi'}{\xi} - 1\right) \geq e_v \qquad \text{for all } v \in S.$$

Hence $\chi(\xi') = \chi(\xi)\chi(\xi'/\xi) = \chi(\xi)$. We can then write

$$\Theta_\chi(D) = \frac{\chi(D)}{q-1} \sum_{\xi \in M_D} \chi(\xi) \cdot \# L(D - F)$$

$$= \frac{\chi(D)}{q-1} \cdot q^{l(D-F)} \sum_{\xi \in M_D} \chi(\xi),$$

where the last sum runs over all elements in the quotient ring

$$M_D = L(D)/L(D - F).$$

If we now observe that $L(D - F) = L(D) \cap A_S(F)$, then the correspondence $f + L(D - F) \to f + A_S(F)$ gives an $\mathbb{F}$-linear inclusion

$$M_D \to A_S/A_S(F) = R,$$

and thus we can think of M_D as an $\mathbb{F}$-submodule of R. We can now write

$$\Theta_\chi(D) = \frac{\chi(D)}{q-1} \cdot q^{l(D-F)}\Lambda_\chi(M_D),$$

where $\Lambda_\chi(M_D)$ is the sum associated with the character χ and the submodule M_D as in 3.5.2. The conditions (3.17), (3.18) imply the primitivity assumptions in (3.11) and (3.12) of Section 3.5.2.

Clearly, as an $\mathbb{F}$-submodule of R, we have

$$\dim_\mathbb{F} M_D = l(D) - l(D - F).$$

The structure of the principal parts of the differential Ω about the points $P_v, v \in S$, implies that the support of the divisor $(\Omega) + F = W + F$ is disjoint from F. Now with a divisor D we associate the divisor

$$D^* = W + F - D$$

and consider as before the $\mathbb{F}$-submodule M_{D^*}.

Claim M_D and M_{D^*} are complementary modules in the sense of Lemma 3.10, i.e. $M_{D^*} = M_D^\perp$.

Proof. To show that two $\mathbb{F}$-submodules M and M' in R are complementary it suffices to show that

(i) $\dim_\mathbb{F} M + \dim_\mathbb{F} M' = \dim_\mathbb{F} R = f$;
(ii) for all $x \in M$ and $y \in M'$, we have $\mathrm{Tr}(xy) = 0$.

From (ii) it follows that M' is an $\mathbb{F}$-submodule of the complementary submodule $M^\perp$ and from (a) and the dimension equality in Lemma 3.10 we get that $M' = M^\perp$. We first prove (a). From the Riemann–Roch Theorem and the equality $\dim_\mathbb{F} M_D = l(D) - l(D - F)$ we obtain

$$\dim_\mathbb{F} M_D + \dim_\mathbb{F} M_{D^*} = \{l(D) - l(D - F)\} + \{l(W + F - D) - l(W - D)\}$$
$$= \{l(D) - l(W - D)\} - \{l(D - F) - l(W + F - D)\}$$
$$= \{\deg D + 1 - g\} - \{\deg(D - F) + 1 - g\}$$
$$= \deg F = f.$$

(ii) Take x an element in M_D and y an element in M_{D^*}, and suppose that ξ and η are respectively pre-images in A_S:

$$A_S \to A_S/A_S(F) \cong R,$$

$$\xi \to x,$$

$$\eta \to y.$$

From the definition of M_D and M_{D^*} we have that the divisors

$$(\xi) + D \tag{3.21}$$

and

$$(\eta) + W + F - D \tag{3.22}$$

are positive; we have also put

$$(\Omega) = W.$$

By adding the last three divisors we obtain that

$$(\xi\eta\Omega) + F \geq 0.$$

Now by the residue theorem we know that the sum of all the residues of the differential $\xi\eta\Omega$ is zero:

$$\sum_v \mathrm{Res}_v(\xi\eta\Omega) = 0,$$

where the sum runs over all the closed points v on C. On the other hand, since $(\xi\eta\Omega) + F \geq 0$, the only points where $\mathrm{Res}_v(\xi\eta\Omega)$ can possibly be non-zero is at the poles of $\xi\eta\Omega$ and these can occur only at the points v in S; therefore we have

$$\sum_{v \in S} \operatorname{Res}_v(\xi\eta\Omega) = 0;$$

this together with Lemma 3.14 proves (b) and we have $M_{D^*} = M_D$.

All of the assumptions in Section 3.5.2 are fulfilled and we can use the formula of Lemma 3.13:

$$\Lambda_{\chi^{-1}}(M_{D^*}) = \|M_D\|^{-1} W(\chi)\Lambda_\chi(M_D). \tag{3.23}$$

If we write Θ for the divisor D^*, we get

$$\Theta_{\chi^{-1}}(D^*) = \frac{\chi^{-1}(D^*)}{q-1} q^{l(W-D)}\Lambda_{\chi^{-1}}(M_{D^*}). \tag{3.24}$$

Substituting (3.23) in (3.24) we obtain

$$\Theta_{\chi^{-1}}(D^*) = \frac{\chi^{-1}(D^*)}{q-1} q^{l(W-D)-l(D)+(D-F)} W(\chi)\Lambda_\chi(M_D)$$

$$= \chi^{-1}(W+F)q^{l(W-D)-l(D)} W(\chi)\Theta_\chi(D). \tag{3.25}$$

From the Riemann–Roch theorem we know that

$$q^{l(W-D)-l(D)} = q^{g-1}ND^{-1} \quad \text{and} \quad NW \cdot NF = q^{2g-2+f}.$$

To obtain the series $L(1 - s, \chi^{-1})$, we multiply the theta relation (3.25) by $(DN^*)^{s-1}$ and use the representation $D^* = W + F - D$ to obtain

$$\Theta_{\chi^{-1}}(D^*)(ND^*)^{-(1-s)} = \chi^{-1}(W+F)W(\chi)q^{g-1+(s-1)(2g-2+f)}\Theta_\chi(D)ND^{-s}. \tag{3.26}$$

Now as D runs over a complete system of absolute divisor classes which are prime to F, so does the divisor $D^* = W + F - D$. Hence if we sum both sides of (3.26) with D running over a complete system of absolute divisor classes and use the representation of $L(s, \chi)$ in terms of $\Theta_\chi(D)$, we finally obtain

Theorem 3.4 (functional equation) *We have*

$$L(1 - s, \chi^{-1}) = \chi^{-1}(W+F)W(\chi)q^{1-g-f} \cdot q^{(2g-2+f)s}L(s, \chi).$$

This functional equation can also be written as

$$L(1 - s, \chi^{-1}) = \varepsilon(\chi)N(W+F)^{s-1/2}L(s, \chi), \tag{3.27a}$$

where we have put

$$\varepsilon(\chi) = \chi^{-1}(W + F)W(\chi)q^{-f/2}.$$

from Remark 1 following Lemma 3.13 we see that $|\varepsilon(\chi)| = 1$.

The case of the divisor $F = \mathcal{O}$ was already considered in Section 3.3. Nevertheless it is instructive to see how the above method can also be applied to this simple case. In this situation we start with a character

$$\chi \colon \mathrm{Div}(C) \to \mathbb{C}$$

with $\mathrm{Div}_a(C) \subset \ker \chi$, i.e. if $D' = D + (f)$, then $\chi(D) = \chi(D')$. We now recall that the absolute divisor class of the divisor D contains $q^{l(D)} - 1$ positive divisors linearly equivalent to d; hence we can write

$$L(s, \chi) = \sum_D \chi(D)ND^{-s}$$

$$= \sum_D{}' \chi(D)(q^{l(D)} - 1)ND^{-s}$$

where the sum $\sum_D$ runs over all positive divisors and $\sum_D'$ runs only over a complete representative system of the absolute divisor classes.

Let us first consider the sum

$$\sum_D{}' \chi(D)ND^{-s};$$

we claim that this sum is equal to 0. In fact if χ is not the trivial character on the group of divisor classes of degree 0, then

$$\sum_{\substack{D \\ \deg D = n}}{}' \chi(D) = 0.$$

On the other hand if $\chi(D) = 1$ for all divisors of degree 0, then letting h denote the number of divisor classes of degree 0 and putting $\chi(D_1) = \varepsilon$ for all divisors of degree 1, we obtain

$$\sum_D{}' \chi(D)ND^{-s} = h \sum_{n = -\infty}^{\infty} \varepsilon^n q^{-ns}.$$

In this last sum, the partial sum $\sum_{n=-\infty}^{k-1}$ converges for $\mathrm{Re}(s) < 0$ and $\sum_{n=k}^{\infty}$ converges for $\mathrm{Re}(s) > 0$ and each gives a rational function in q^s and the whole series, which is the sum of the two functions is 0, because it remains invariant when multiplied by εq^{-s}.

Adding the two sums above we obtain

$$L(s, \chi) = \sum_D{}' \chi(D)q^{l(D)}q^{-s \deg D}.$$

Let w represent a divisor in the canonical class. As D runs over a complete representative system for the divisor classes in $\mathrm{Div}(C)$, so does $w - D$;

substituting $w - D$, χ^{-1} and $1 - s$ in the last expression for D, χ and s we obtain

$$L(1 - s, \chi^{-1}) = \sum_D{}' \chi^{-1}(w - D)q^{l(w-D)}q^{(s-1)(\deg W - \deg D)}.$$

On the other hand the Riemann–Roch theorem gives

$$l(w - D) - (\deg w - \deg D) = l(D) - g + 1;$$

Therefore we have

$$L(1 - s, \chi^{-1}) = \sum_D{}' \chi(D)\chi^{-1}(W)q^{s \deg W}\frac{q^{l(D)-g+1}}{q^{s \deg D}}$$

$$= \chi^{-1}(W)q^{1-g}q^{(2g-2)s}L(s, \chi).$$

This proves the functional equation in the case $D = \mathcal{O}$; we observe that when $F = \mathcal{O}$, $f = 0$ and $w(\chi) = 1$.

Remark 1. If D is a divisor with $\deg D < 0$, then the vector space $L(D) = 0$. If D is a divisor with $\deg D > 2g - 2 + f$, it follows that the divisor $D^* = W + F - D$ has $\deg D^* < 0$ and therefore $L(D^*) = 0$. From the representation

$$\Theta_{\chi^{-1}}(D^*) = \frac{\chi^{-1}(D^*)}{q - 1} \sum_{\xi \in L(D^*)} \chi^{-1}(\xi)$$

and the theta relation (3.25) it follows that if $\deg D > 2g - 2 + f$, then $\Theta_\chi(D) = 0$. If we make the substitution $t = q^{-s}$ in the representation

$$L(s, \chi) = \sum_D \Theta_\chi(D)ND^{-s}$$

we obtain that the resulting series $L(t, \chi)$ is actually a polynomial of degree $\leq 2g - 2 + f$. Making the substitution $q^{-s} = t$ in the functional equation (3.27a) we obtain

$$L\left(\frac{1}{qt}, \chi^{-1}\right) = \varepsilon(\chi)(\sqrt{q}/t)^{2g-2+f}L(t, \chi). \tag{3.27b}$$

Comparing the degrees on both sides we obtain that

$$\deg L(t, \chi) = 2g - 2 + f.$$

2. The deeper significance of the last result is to be found in Grothendieck's cohomological interpretation of L-functions. In fact, given χ, there is an l-adic sheaf $\mathscr{L}_\chi$ on the curve $\bar{C} = C \times F$ and there are l-adic cohomology groups $V_l = H^1(\bar{C}, \mathscr{L}_\chi)$ of dimension $2g - 2 + f$ on which the Frobenius automorphism F acts by transport of structure. Furthermore

$$L(t, \chi) = \det(1 - Ft | V_l).$$

This equality together with the functional equation imply that

$$\varepsilon(\chi)(\sqrt{q})^{2g-2+f} = \det(-F/V_l)^{-1}$$

(Deligne [14], p. 190).

Exercises

1. Prove that if Z is the compositum of two disjoint Abelian extensions K and K' of the rational function field $K_0 = \mathbb{F}_q(x)$ then the following relation holds between their L-functions:

$$\zeta_K(s) = \prod_\chi L(s, \chi),$$

$$\zeta_{K'}(s) = \prod_\psi L(s, \psi),$$

$\zeta_Z(s) = \prod_{\chi, \psi} L(s, \chi\psi)$, where χ (resp. ψ) runs over all the characters of $\mathrm{Gal}(K/K_0)$ (resp. $\mathrm{Gal}(K'/K_0)$).

2. Let ψ be a character of $\mathrm{Gal}(K'/K_0)$ and for a divisor A of K put $\psi(A) = \psi(N_{K/K_0}(A))$. Prove that

$$L(s, \psi, z/K) = \prod_\chi L(s, \psi \circ \chi, K/K_0),$$

where χ runs over all characters of $\mathrm{Gal}(K/K_0)$.

3. Suppose $\mathbb{F}_q$ contains an n-th root of 1 and let $K = K_0(y)$ be the Kummer extension of $K_0 = \mathbb{F}_q(x)$ generated by the roots of $y^n = b(x)$, where $b(x)$ is a function in $\mathbb{F}_q(x)$. Define a character $\chi: \mathbb{F}_q^\times \to C^\times$ by $\chi(a) = e^{2\pi i v/n}$, where $a^{(q-1)/n} = \zeta^v$. Let P be a point distinct from the support of the principal divisor $(b(x))$; let $N(P) = q^{\deg(P)}$ and define for any $z \in K$, the Frobenius substitution σ_P by

$$z^{N(P)} \equiv \sigma_P(z) \bmod P$$

Prove that

$$\sigma_P(y)y^{-1} \equiv (Nb(x))^{(q-1)/n} \bmod P, \quad Nb(x) = b(x)^{1+q+\cdots+q^{d-1}}, \quad d = \deg P.$$

Hence define the n-th power residue symbol by

$$\left(\frac{b(x)}{P}\right)_n = \chi(Nb(x)).$$

Since $\mathbb{F}_q(x)$ can be thought of as the function field of the projective line $\mathbb{P}^1$, we obtain a character

$$\left(\frac{b(x)}{*}\right)_n : \mathrm{Div}(\mathbb{P}^1 - \mathrm{Supp}(b(x))) \to C^\times$$

by setting for $D = \sum_P m_P P$, $(b(x)/D)_n = \prod_P (b(x)/P)_n^{m_P}$.

4. Prove that if $\deg(P) = 1$, so that the principal divisor $(x - a) = P - P_\infty$ with an $a \in \mathbb{F}_q$, then

$$\left(\frac{b(x)}{P}\right)_n = \chi(b(a)).$$

5. (RECIPROCITY LAW) Let $a(x) = \prod_{i=1}^m (x - \alpha_i)$, $b(x) = b \prod_{j=1}^l (x - \beta_j)$, $b \in \mathbb{F}_q^\times$, be polynomials whose associated principal divisors are $(a(x)) = A - mP_\infty$, $(b(x)) = B - lP_\infty$, so that A and B are positive. Show that

$$\left(\frac{b(x)}{A}\right)_n = \chi(-1)^{ml}\chi(b)^m\left(\frac{a(x)}{B}\right)_n.$$

6. Let m and n be positive integers with $n|(q-1)$ and $m|(q-1)$. Let ζ be a generator of $\mathbb{F}_q^\times$ and let ω be a primitive $(q-1)$ root of 1. We define two characters χ, ψ: $\mathbb{F}_q^\times \to \mathbb{C}^\times$ by putting $\chi(a) = \omega^{v((q-1)/n)}$, $\psi(a) = \omega^{v((q-1)/n)}$, when $a = \zeta^v$.

6a. Let $K = K_0(x)$ be the Kummer extension obtained from the rational function field $K_0 = \mathbb{F}_q(t)$ by adjoining a root of the equation $x^m = t$. Show that all the characters of $\mathrm{Gal}(K/K_0)$ are given by the m-th power residue symbols

$$\left(\frac{t^\mu}{A}\right)_m, \mu = 0, \ldots, m - 1,$$

constructed with the characters χ^μ: $\mathbb{F}_q^\times \to \mathbb{C}^\times$, and that the conductor is $\mathcal{F}_\mu = P_0 + P_\infty$, except when $\mu = 0$ in which case $\mathcal{F}_\mu = (1)$; the principal divisor $(t) = P_0 + P_\infty$.

6b. Let $K' = K_0(y)$ be the Kummer extension obtained from the rational function field $K_0 = \mathbb{F}_q(t)$ by adjoining a root of the equation $y^n = 1 - t$. Show that all the characters of $\mathrm{Gal}(K'/K_0)$ are given by the n-th power residue symbols

$$\left(\frac{(1-t)^v}{A}\right)_n, v = 0, \ldots, n - 1,$$

constructed with the characters ψ^v: $\mathbb{F}_q^\times \to \mathbb{C}^\times$ and that the conductor is $\mathcal{F}_v' = P_1 + P_\infty$; except when $v = 0$ in which case $\mathcal{F}_v' = (1)$; the principal divisor $(1 - t) = P_1 + P_\infty$.

6c. Let KK' be the compositum of K and K' and use the following property of the power residue symbol

$$\left(\frac{A}{B}\right)_l = \left(\frac{B^{l'}}{A}\right)_w$$

to show that all the characters of $\mathrm{Gal}(KK'/K_0)$ are given by

$$\left(\frac{t^\mu}{A}\right)_m\left(\frac{(1-t)^v}{A}\right)_n = \left(\frac{t^{\mu n_0}}{A}\right)_l\left(\frac{(1-t)^{v m_0}}{A}\right)_l = \left(\frac{t^{\mu n_0}(1-t)^{v m_0}}{A}\right)_l,$$

where $l = \mathrm{l.c.d.}(m,n)$, $m = dm_0$, $n = dn_0$, $d = \mathrm{g.c.d.}(m,n)$, and $\mu = 0, \ldots, m - 1$, $v = 0, \ldots, n - 1$. Also show that the conductors of these characters are given by

$$(1), \mu = 0, \nu = 0$$

$$P_0 + P_\infty, \mu = 1, \ldots, m - 1, \nu = 0$$

$$P_1 + P_\infty, \mu = 0, \nu = 1, \ldots, n - 1$$

$$P_0 + P_1, \mu = \delta m_0, \nu = (d - \delta)n_0, \delta = 1, \ldots, d - 1$$

and

$$P_0 + P_1 + P_2$$

for all other pairs μ, ν.

6d. Define the L-functions

$$L_{\mu\nu}(s) = \sum_A \left(\frac{t^{\mu n_0}(1 - t)^{\nu m_0}}{A} \right)_t,$$

$\mu = 0, \ldots, m - 1, \nu = 0, \ldots, n - 1$. Use the fact that the rational function field $\mathbb{F}_q(t)$ is of genus 0 to show that

(i) $\qquad\qquad L_{00}(s) = \zeta_{K_0}(s) = \zeta_0(s);$

(ii) $\qquad\qquad L_{\mu 0}(s) = 1, \mu = 1, \ldots, m - 1;$

(iii) $\qquad\qquad L_{0\nu} = 1, \nu = 1, \ldots, n - 1;$

(iv) $\qquad\qquad L_{\delta m_0, (d-\delta)n_0}(s) = 1, \delta = 1, \ldots, d - 1.$

(v) $\qquad\qquad L_{\mu\nu}(s) = 1 - \pi(\chi^\mu, \psi^{n u})q^{-s},$

for all other pairs (μ, ν), where $\pi(\chi^\mu, \psi^\nu)$ is a complex number of absolute value $q^{1/2}$.

(Hint: For (ii), (iii), (iv) use the fact that the degree of $L_{\mu\nu}(s)$ as a polynomial in q^{-s} is $\deg \mathscr{F}_{\mu\nu} - 2$.)

6e. Show that the zeta function of the compositum KK' is

$$\zeta_{KK'}(s) = \prod_{\mu=0}^{m-1} \prod_{\nu=0}^{n-1} L_{\mu\nu}(s),$$

and that the non-trivial part of $\zeta_{KK'}(s)$ is given by

$$L_{KK'} = \frac{\zeta_{KK'}(s)}{\zeta_0(s)} = \prod_{\mu, nu}^{*} L_{\mu, \nu}(s),$$

where (μ, ν) ranges over all pairs of integers $\mu = 1, \ldots, m - 1, \nu = 1, \ldots, n - 1$, satisfying the condition $\chi^\mu \neq 1, \psi^\nu \neq 1$ and $\chi^\mu\psi^\nu \neq 1$.

Notes

Weil gave two distinct proofs of the Riemann hypothesis for function fields over finite fields. The first is a geometric argument based on the theory of correspondences and ultimately depends on algebraic varieties of dimension greater than one. The second is based on the theory of abelian varieties and uses 1-adic vector spaces modeled on the classical homology groups of algebraic curves over the complex numbers. Both of these proofs were originally difficult to study because they were presented in the language of Weil's Foundations. The first real break-through in simplifying the proof came about when Bombieri realized that some preliminary cases dealt with by Stepanov could actually be generalized to give the full proof of the Riemann hypothesis using only a weak form of the Riemann–Roch theorem. This is the proof that we have presented in this chapter. Recently Voloch and Stohr (*Proc. London Math. Soc.*, **52** (1986)) have given a new proof based on the theory of Weierstrass points and obtain several improvements. The deeper understanding of the problems considered in this chapter eventually leads to a study of Deligne's proof of the Weil conjectures and their arithmetic consequences.

4
Exponential sums

Many interesting exponential sums which arise naturally in number theory give rise to L-functions of algebraic curves. This connection is made explicit in this chapter by means of several examples related to Gauss and Kloosterman sums. The results of Chapter 3 are also used to obtain bounds for exponential sums.

4.1 The zeta function of the projective line

Let $\mathbb{F}_q$ be the finite field of q elements and let $A = \mathbb{F}_q[x]$ be the ring of polynomials with coefficients in $\mathbb{F}_q$. The set of closed points on the projective line $\mathbb{P}^1$ can be identified with the set of monic irreducible polynomials in A plus the rational function x^{-1} which corresponds to the point at infinity on $\mathbb{P}^1$. If P is a polynomial in A of degree d, we put $NP = q^{-d}$. The zeta function of the affine line $\mathbb{A}^1 = \mathbb{P}^1 - \{\infty\}$ is defined, for s a complex number, by

$$Z(s, \mathbb{A}^1) = \sum_a Na^{-s},$$

where a runs over all monic polynomials in A including $a = 1$. Since the number of distinct monic polynomials in A of degree n is q^n, we have

$$Z(s, \mathbb{A}^1) = \sum_{n=0}^{\infty} q^n q^{-ns} = \frac{1}{1 - q^{1-s}};$$

hence $Z(s, \mathbb{A}^1)$ is defined by a series which converges absolutely for $\mathrm{Re}(s) > 1$. Furthermore since A is a unique factorization domain, we have an Euler product expansion

$$Z(s, \mathbb{A}^1) = \prod_P \frac{1}{1 - NP^{-s}},$$

where P runs over all monic irreducible polynomials in A of degree ≥ 1. If we include in this Euler product the factor $(1 - q^{-s})^{-1}$ which corresponds to the rational function $P_\infty = 1/x$, we obtain the zeta function of the projective line

$$Z(s, \mathbb{P}^1) = \prod_P \frac{1}{1 - NP^{-s}} = \frac{1}{1 - q^{-s}} \cdot \frac{1}{1 - q^{1-s}}.$$

To study $Z(s, \mathbb{A}^1)$ we can also proceed in a different way. First we recall that a fundamental result in the arithmetic of the ring A is the theorem of Gauss to the effect that for any positive integer $n \geq 1$

$$x^{q^n} - x = \prod_{d|n} F_d(x),$$

where $F_d(x)$ is the product of all monic irreducible polynomials in A of degree d. By comparing the degrees on both sides of this identity we obtain

$$q^n = \sum_{d|n} dN_d,$$

where N_d is the number of monic irreducible polynomials in A of degree d. In the Euler product for $Z(s, \mathbb{A}^1)$ we collect those polynomials P of degree d and use the above expression for q^n to obtain

$$Z(s, \mathbb{A}^1) = \prod_{d=1}^{\infty} \left\{ \frac{1}{1 - q^{-ds}} \right\}^{N_d}.$$

On taking the logarithm of both sides we obtain

$$\log Z(s, \mathbb{A}^1) = \sum_{d=1}^{\infty} N_d \sum_{k=1}^{\infty} q^{-sdk}/k$$

$$= \sum_{m=1}^{\infty} \frac{1}{m} q^{-sm} \sum_{d|m} dN_d$$

$$= \sum_{m=1}^{\infty} \frac{1}{m} (q^{1-s})^m = \log \frac{1}{1 - q^{1-s}};$$

this agrees with the expression obtained earlier for $Z(s, \mathbb{A}^1)$. The following observations are simple consequences of the above representation:

$$Z(s, \mathbb{P}^1) = Z(1 - s, \mathbb{P}^1), \tag{4.1}$$

$Z(s, \mathbb{P}^1)$ has a meromorphic continuation to the whole s-plane; it has simple poles at $s = 0, 1$, (4.2)

the product expansion of $Z(s, \mathbb{P}^1)$ contains an infinite number of local factors (Euler's proof of the infinitude of primes!) (4.3)

$Z(1 + it, \mathbb{P}^1) \neq 0$ for all real t. (4.4)

4.2 Gauss sums: first example of an *L*-function for the projective line

If x is a complex number and if m is an integer ≥ 1, we put

$$e_m(x) = e^{2\pi i x/m}.$$

Let p denote a prime number. If $x \in \mathbb{Z}$, and μ_p denotes the group of p-th roots of unity, then the map $x \to e_p(x)$ defines by periodicity an isomorphism

$$e_p : \mathbb{Z}/p\mathbb{Z} \to \mu_p.$$

Let $k = \mathbb{F}_q$ denote the finite field with $q = p^a$ elements. For $x \in \mathbb{F}_q$ we put $\psi(x) = e_p(\mathrm{Tr}_{R/\mathbb{F}_p}(x))$. Let $g(\chi, \psi)$ be the gauss sum defined with ψ and a multiplicative character χ (see §4.3). Note that $|g(\chi, \psi)|^2 = q$, for any non-trivial χ.

To define an *L*-function related to $g(\chi, \psi)$ we proceed as follows. Let

$$a = x^n + a_1 x^{n-1} + \cdots + a_n$$

be a monic polynomial in the ring $A = \mathbb{F}_q[x]$ and with characters χ and ψ as above, define a function on A by

$$\Lambda(a) = \chi(a_n)\psi(a_1).$$

Observe that if

$$b = x^m + b_1 x^{m-1} + \cdots + b_m$$

is another polynomial, then

$$a \cdot b = x^{m+n} + (a_1 + b_1)x^{m+n-1} + \cdots + a_n b_m,$$

and hence

$$\Lambda(a \cdot b) = \Lambda(a)\Lambda(b).$$

We can thus form the zeta function

$$Z(s, \Lambda, \mathbb{P}^1) = \sum_a \Lambda(a)Na^{-s} = \prod_P \frac{1}{1 - \Lambda(P)NP^{-s}},$$

where the product runs over all irreducible monic polynomials in A. From the properties of $Z(s, \mathbb{P}^1)$ it follows easily that $Z(s, \Lambda, \mathbb{P}^1)$ converges absolutely for $\mathrm{Re}(s) > 1$. Now, the Dirichlet series $Z(s, \Lambda, \mathbb{P}^1)$ is also expressible in the form

$$Z(s, \Lambda, \mathbb{P}^1) = 1 + \sum_{d=1}^{\infty} q^{-ds} S_d,$$

where

$$S_d = \sum_a \Lambda(a)$$

and the sum runs over all monic polynomials of degree d. As all monic polynomials of degree 1 in A are of the form $a = x + c$ with $c \in \mathbb{F}_q$, and since $\Lambda(x + c) = \chi(c)\psi(c)$, we obtain for $d = 1$ the Gauss sum $S_1 = g(\chi, \psi)$. Also all monic polynomials in A of degree 2 have the form $a = x^2 + bx + c$ with $b, c \in \mathbb{F}_q$; for these we have

$$S_2 = \sum_a \Lambda(x^2 + bx + c) = \sum_b \sum_c \chi(c)\psi(b) = \sum_c \chi(c)\left(\sum_b \psi(b)\right) = 0.$$

A similar argument shows that for all $d \geq 3$, $S_d = 0$. Hence we obtain

$$Z(s, \Lambda, \mathbb{P}^1) = 1 + g(\chi, \psi)q^{-s}.$$

This representation implies that $Z(s, \Lambda, \mathbb{P}^1)$, originally defined only for $\mathrm{Re}(s) > 1$, has an analytic continuation to the whole s-plane; the fact $|g(\chi, \psi)| = q^{1/2}$ implies that the zeros of $Z(s, \Lambda, \mathbb{P}^1)$ are all located on the line $\mathrm{Re}(s) = 1/2$. The weaker statement $|g(\chi, \psi)| < q$ would suffice to show that $Z(1 + it, \Lambda, \mathbb{P}^1) \neq 0$ for all real t.

4.3 Properties of Gauss sums

4.3.0 Cyclotomic extensions: basic facts

For a positive integer m, denote by μ_m the group of all roots of $X^m = 1$; for a prime p, μ_p is defined similarly. We define $\mathbb{Q}(\mu_m)$ to be the field obtained by adjoining the elements of μ_m to the rationals; this is also the field obtained by adjoining a primitive root ζ_m of $X^m = 1$; $\mathbb{Q}(\mu_p)$ and $\mathbb{Q}(\mu_{mp})$ are defined in the same way. We denote by $\mathcal{O}$ the ring of integers of the field $\mathbb{Q}(\mu_m)$, i.e. $\mathcal{O} = \mathbb{Z}[\zeta_m]$. The irreducibility of the cyclotomic equation implies that the degree of $\mathbb{Q}(\mu_m)$ over $\mathbb{Q}$ is $\varphi(m)$; this in turn yields a bijection

$$(\mathbb{Z}/m\mathbb{Z})^\times \xrightarrow{\sim} \mathrm{Gal}(\mathbb{Q}(\mu_m)/\mathbb{Q}), \qquad u \mapsto \sigma_u,$$

between the residue classes relatively prime to m and the Galois group of $\mathbb{Q}(\mu_m)$ over $\mathbb{Q}$, where the automorphism σ_u acts via the map $\zeta_m \mapsto \zeta_m^u$. If m and m' are relatively prime, then the compositum of $\mathbb{Q}(\mu_m)$ and $\mathbb{Q}(\mu_{m'})$ is $\mathbb{Q}(\mu_{mm'})$; also $\mathbb{Q}(\mu_m) \cap \mathbb{Q}(\mu_{m'}) = \mathbb{Q}$ and therefore the Galois group of $\mathbb{Q}(\mu_{mm'})$ over $\mathbb{Q}$ is the direct product of those of $\mathbb{Q}(\mu_m)$ and $\mathbb{Q}(\mu_{m'})$ over $\mathbb{Q}$. In particular we have a diagram of field extensions

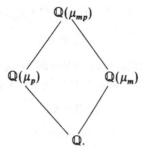

Since p does not divide m, the prime ideal (p) does not ramify in $\mathbb{Q}(\mu_m)$. Let f be the order of p in $(\mathbb{Z}/m\mathbb{Z})^\times$ (i.e. the smallest positive integer such that $p^f \equiv 1 \bmod m$). If $\mathcal{Y}$ is a prime ideal of $\mathbb{Q}(\mu_m)$ which contains the rational prime p, that is a prime ideal of $\mathbb{Q}(\mu_m)$ which lies above the prime ideal (p) of $\mathbb{Q}$, then $\mathcal{O}/\mathcal{Y} = \mathbb{F}_q$ the finite field with $q = p^f$ elements. Let

$$D = \{\sigma \in \mathrm{Gal}(\mathbb{Q}(\mu_m)/\mathbb{Q}) \colon \mathcal{Y}^\sigma = \mathcal{Y}\}$$

be the decomposition group of $\mathcal{Y}$; D is isomorphic to the subgroup of $(\mathbb{Z}/m\mathbb{Z})^\times$ generated by the residue class $\bar{p}$ of p. If

$$\mathrm{Gal}(\mathbb{Q}(\mu_m)/\mathbb{Q}) = \bigcup_{j=1}^{g} \sigma_{u_j} D$$

is a coset decomposition, then the g prime ideals $\mathcal{Y}^{\sigma_{u_j}}$ are all distinct and divide (p) exactly to the first power:

$$(p) = \mathcal{Y}^{\sigma_{u_1}} \ldots \mathcal{Y}^{\sigma_{u_g}}.$$

The extension $\mathbb{Q}(\mu_p)$ is of degree $p - 1$ over $\mathbb{Q}$. The only rational prime which ramifies in $\mathbb{Q}(\mu_p)$ is p; in fact, if $\pi = \zeta_p - 1$, then $\mathcal{Y}' = (\pi)$ is a prime ideal of $\mathbb{Q}(\zeta_p)$ and $(p) = (\mathcal{Y}')^{p-1}$. Let $\mathcal{F}$ be a prime ideal of $\mathbb{Q}(\mu_{mp})$ such that $\mathcal{Y} = \mathcal{O} \cap \mathcal{F}$. Let $\mathcal{Y}' = \mathcal{O}_{\mathbb{Q}(\mu_p)} \cap \mathcal{F}$; $\mathcal{Y}'$ does not ramify in $\mathbb{Q}(\mu_{mp})$ and $\mathcal{F} \| \mathcal{Y}'$ and $\mathcal{Y} = \mathcal{F}^{p-1}$.

The image of an element $\mathscr{E} \in \mathcal{O}$ under the reduction map

$$\mathcal{O} \to \mathcal{O}/\mathcal{Y} = \mathbb{F}_q$$

is denoted by $\bar{\xi}$. Let ζ_m be a primitive m-th root of 1, i.e. a generator of the cyclic group μ_m. Let $\bar{\mu}_m$ be the subgroup of $\mathbb{F}_q^\times$ generated by $\bar{\zeta}_m$. Since $(p, m) = 1$ the map $\varphi \colon \mu_m \to \bar{\mu}_m$ is a bijection, for otherwise the norm of $\zeta_m - 1$ would be divisible by p and this can be seen to be impossible by using the m-th cyclotomic equation; therefore μ_m and $\bar{\mu}_m$ have the same order m which divides $q - 1$; the integer $(q - 1)/m$ is denoted by n. Since $\mathbb{F}_q^\times$ is cyclic of order mn, $\bar{\mu}_m = (\mathbb{F}_q^\times)^n$.

If $\xi \in \mathcal{O}$ and $\xi \notin \mathcal{Y}$, then $\bar{\xi} \in \mathbb{F}_q^\times$. Hence $\bar{\xi}^n \in \bar{\mu}_m$ and

$$\bar{\xi}^n \equiv \zeta_m^a \bmod \mathcal{Y}$$

for some $a \in \mathbb{Z}$. Since $\mu_m \xrightarrow{\sim} \bar{\mu}_m$ is an isomorphism, ζ_m^a is uniquely determined by ξ. We define a power residue symbol by putting

$$(\xi/\mathcal{Y}) = \zeta_m^a.$$

The following properties are readily verified:

(i) $(\xi/\mathcal{Y})$ depends only on the residue class $\bar{\xi}$,
(ii) $(\xi\xi'/\mathcal{Y}) = (\xi/\mathcal{Y})(\xi'/\mathcal{Y})$,
(iii) $\{(\xi/\mathcal{Y}): \xi \in \mathcal{O}, \xi \notin \mathcal{Y}\} = \mu_m$.
(iv) $(\xi/\mathcal{Y}) \equiv \xi^n \bmod \mathcal{Y}, n = (q-1)/m$.

For $x \in \mathbb{F}_q^\times$ we define $\chi(x)$ to be $(\xi/\mathcal{Y})$ for some $\xi \in \mathcal{O}$ such that $\bar{\xi} = x$. χ defines a character of the multiplicative group $\mathbb{F}_q^\times$ of order m:

$$\chi: \mathbb{F}_q^\times \to \mu_m.$$

Let $v = \mathrm{ord}_p$ be the p-adic exponential valuation of $\mathbb{Q}$, i.e. $v(p) = 1$ and $v(a \cdot b) = v(a) + v(b)$. The completion of $\mathbb{Q}$ with respect to v is denoted by $\mathbb{Q}_p$. We also denote by v the unique extension of v from $\mathbb{Q}_p$ to $\mathbb{Q}_p(\mu_p)$ so that, since $(p) = (\mathcal{Y}')^{p-1}$ in $\mathbb{Q}(\mu_p)$, we have $v(p) = (p-1)v(\pi)$ and hence $v(\pi) = 1/(p-1)$.

The group $\mathrm{Gal}(\mathbb{F}_q/\mathbb{F}_p)$ is cyclic of order f and is generated by the Frobenius substitution

$$\sigma(x) = x^p, \qquad x \in \mathbb{F}_q;$$

hence the trace and norm from $\mathbb{F}_q$ to $\mathbb{F}_p$ are given by

$$T(x) = x + x^p + \cdots + x^{p^{f-1}}$$

$$N(x) = x^{1+p+\cdots+p^{f-1}} = x^{(p^f-1)/(p-1)}.$$

Clearly $T(x + y) = T(x) + T(y)$.

Let ε be a primitive root of $X^p = 1$, i.e. a generator of μ_p. Identifying $\mathbb{F}_p$ with the integers modulo p, we can define as in Section 4.2 a character of the additive group $\mathbb{F}_q$:

$$\psi: \mathbb{F}_q^+ \to \mu_p, \qquad \psi(x) = \zeta_p^{T(x)}.$$

Since the trace map T is onto, clearly $\psi \neq 1$.

Let $\sigma \in \mathrm{Gal}(\mathbb{Q}(\mu_m)/\mathbb{Q})$ be such that $\sigma(\zeta_m) = \zeta_m^s$, s a positive integer prime to m. Since $\mathrm{Gal}(\mathbb{Q}(\mu_{mp})/\mathbb{Q})$ is the direct product of $\mathrm{Gal}(\mathbb{Q}(\mu_m)/\mathbb{Q})$ and $\mathrm{Gal}(\mathbb{Q}(\mu_p)/\mathbb{Q})$, there is a unique extension of σ to an automorphism in

$\text{Gal}(\mathbb{Q}(\mu_{mp})/\mathbb{Q})$ which leaves $Q(\mu_p)$ pointwise fixed. In particular $\psi^\sigma = \psi$ and $\chi^\sigma = \chi^s$.

4.3.1 Elementary properties

In this section we establish those properties of $G(\chi, \psi)$ which follow easily from the definition.

Proposition 4.1 (i) $|G(\chi, \psi)|^2 = q$ *if* $\chi \neq 1, \psi \neq 1$.

$$\text{(i)} \quad G(\chi, \psi) = \begin{cases} q - 1 & \text{if } \chi = 1, \psi = 1 \\ 0 & \text{if } \chi \neq 1, \psi = 1. \\ -1 & \text{if } \chi = 1, \psi \neq 1 \end{cases}$$

Proof. The second assertion is a simple consequence of the orthogonality relations for the character group of $\mathbb{F}_q^+$ or $\mathbb{F}_q^\times$. As for the first assertion we have

$$|G(\chi, \psi)|^2 = G(\chi, \psi)\overline{G(\chi, \psi)}$$

$$= \sum_{x \neq 0} \sum_{y \neq 0} \chi(x)\psi(x)\overline{\chi}(y)\overline{\psi}(y)$$

$$= \sum_{x \neq 0} \sum_{y \neq 0} \chi(x/y)\psi(x - y),$$

where we have used the fact that $\overline{\psi}(y) = \psi(y)^{-1} = \psi(-y)$ and $\overline{\chi}(y) = \chi(y)^{-1} = \chi(y^{-1})$. Hence the above double sum is equal to

$$= \sum_{y, z \neq 0} \psi((z - 1)y)\chi(z), \qquad (z = x/y).$$

First we compute $\sum_{y \neq 0} \psi((z - 1)y)$. We need the basic fact that if $\psi: \mathbb{F}_q^+ \to \mathbb{C}$ is a non-trivial character of the additive group, then any other character of $\mathbb{F}_q^+$ is of the form $\psi_c(x) = \psi(cx)$ for some $c \in \mathbb{F}_q$. Observe that $\psi_c(x) = 1$ if and only if $c = 0$. Thus

$$\sum_{y \in \mathbb{F}_q} \psi_c(y) = \begin{cases} q & \text{if } c = 0 \\ 0 & \text{if } c \neq 0. \end{cases}$$

Hence

$$\sum_{y \neq 0} \psi_c(y) = \begin{cases} q - 1 & \text{if } c = 1 \\ -1 & \text{if } c \neq 1 \end{cases}$$

and

$$\sum_{y \neq 0} \psi_c((z-1)y) = \begin{cases} q-1 & \text{if } z = 1 \\ -1 & \text{if } z \neq 1. \end{cases}$$

Therefore

$$|G(\chi, \psi)|^2 = \chi(1)(q-1) + \sum_{\substack{z \neq 0 \\ z \neq 1}} (-1)\chi(x)$$

$$= \chi(1)q - \sum_{z \neq 0} \chi(z)$$

$$= q,$$

since $\chi \neq 1$. This proves the proposition.

Proposition 4.2 Let $\psi: \mathbb{F}_q^+ \to \mathbb{C}^\times$ *be a non-trivial additive character and* $\chi: \mathbb{F}_q^\times \to \mathbb{C}^\times$ *a non-trivial multiplicative character of order m. Then*

(i) $G(\chi, \psi)$ *is an algebraic integer in* $\mathbb{Q}(\chi, \psi)$,
(ii) $G(\chi, \psi)^m$ *is an algebraic integer in* $\mathbb{Q}(\chi)$,
(iii) $G(\chi, \psi)^m = G(\chi, \psi')^m$ *for any other non-trivial character* $\psi' \neq 1$,
(iv) $G(\chi, \psi)^m$ *and* $G(\chi', \psi)^m$ *are conjugate in* $\mathbb{Q}(\chi)$ *if* χ' *is another character of the same order m.*

Proof. (i) is clear since $G(\chi, \psi)$ is a sum of roots 1 in $Q(\chi, \psi)$. As for (iii), we have already observed that $\psi' = \psi_c$ for some $c \in \mathbb{F}_q^\times$. Therefore

$$G(\chi, \psi') = \sum_{x \in \mathbb{F}_q^\times} \chi(x)\psi(cx)$$

$$= \sum_{y \neq 0} \chi(y/c)\psi(y), \qquad (y = cx),$$

$$= \sum_{y \neq 0} \chi(c)^{-1}\chi(y)\psi(y)$$

$$= \chi(c)^{-1}G(\chi, \psi).$$

Since $\chi(c)$ is an m-th root of 1, (iii) follows. In the proof of (ii) we consider an automorphism $\sigma \in \text{Gal}(\mathbb{Q}(\chi, \psi)/\mathbb{Q}(\chi))$ and define new characters $\psi^\sigma: \mathbb{F}_q^+ \to \mathbb{C}^+$ by $\psi^\sigma(x) = (\psi(x))^\sigma$ and $\chi^\sigma: \mathbb{F}_q^\times \to \mathbb{C}^\times$ by $\chi^\sigma(x) = (\chi(x))^\sigma$. These are both non-trivial characters and clearly $\chi^\sigma = \chi$ since σ leaves the value of χ fixed. Now, by (iii) we have

$$(G(\chi, \psi)^m)^\sigma = G(\chi, \psi^\sigma)^m = G(\chi, \psi)^m.$$

Hence $G(\chi, \psi)^m$ is left fixed by all of $\text{Gal}(\mathbb{Q}(\chi, \psi)/\mathbb{Q}(\chi))$ and must therefore be an element of $\mathbb{Q}(\chi)$. It is clearly an algebraic integer.

(iv) If $\chi': \mathbb{F}_q^\times \to \mathbb{C}^\times$ is another character of order m, we can find, using the fact that $\mathbb{F}_q^\times$ and its character group are cyclic, an integer s prime to m

such that $\chi' = \chi^s$. Let $\sigma \in \text{Gal}(\mathbb{Q}(\zeta_m)/\mathbb{Q})$ be such that $\sigma(\zeta_m) = \zeta_m^s$. From the Galois theory of cyclotomic extensions we know that σ has a unique extension to an automorphism (also denoted by σ) of $\mathbb{Q}(\chi, \psi)$ which acts trivially on $\mathbb{Q}(\psi)$. For this σ we have $\psi^\sigma = \psi$ and $\chi^\sigma = \chi^s$. Hence

$$G(\chi, \psi)^\sigma = G(\chi^\sigma, \psi^\sigma) = G(\chi^s, \psi) = G(\chi', \psi).$$

Thus $G(\chi, \psi)$ and $G(\chi', \psi)$ are conjugate by an element $\sigma \in \text{Gal}(\mathbb{Q}(\chi, \psi)/\mathbb{Q})$; the m-th powers also lie in $\mathbb{Q}(\chi)$ and are conjugate under the restriction of σ to $\mathbb{Q}(\chi)$. This proves the proposition.

4.3.2 The Hasse–Davenport relation

Let $F = \mathbb{F}_q$ be the finite field with q elements and let $F_s = \mathbb{F}_{q^s}$ be the extension of F of degree s. If χ (resp. ψ) is a multiplicative (resp. additive) character of F, we define characters F_s by putting

$$\chi' = \chi \circ N_{F_s/F}, \qquad \psi' = \psi \circ t_{F_s/F}.$$

Theorem 4.1 (Hasse–Davenport relation) *With assumptions and notations as above we have*

$$-G(\chi', \psi') = (-G(\chi, \psi))^s.$$

We first establish the following two elementary lemmas.

Lemma 4.1 *Let $\alpha \in F_s$ and let $f(x)$ be the monic irreducible polynomial for α over F of degree d. Then $\Lambda(f)^{s/d} = \chi'(\alpha)\psi'(\alpha)$.*

Proof. Recall from Section 4.2 that if $f = x^d - c_1 x^{d-1} + \cdots + (-1)^d c_d$ then $\Lambda(f) = \chi(c_d)\psi(c_1)$; $F_d = F(\alpha)$ is an extension of F of degree d, and hence F_s is of degree s/d over F_d. Now

$$N_{F_s/F}(\alpha) = N_{F_d/F}(N_{F_s/F_d}(\alpha)) = N_{F_d/F}(\alpha)^{s/d} = c_d^{s/d}$$

and

$$\text{Tr}_{F_s/F}(\alpha) = \text{Tr}_{F_d/F}(\text{Tr}_{F_s/F_d}(\alpha)) = \text{Tr}_{F_d/F}(s^\alpha/d) = c_1(s/d);$$

therefore

$$\chi'(\alpha)\psi'(\alpha) = \chi(c_d)^{s/d}\psi(c_1)^{s/d} = \Lambda(f)^{s/d}.$$

Lemma 4.2 *We have*

$$G(\chi', \psi') = \sum_f (\deg f)\Lambda(f)^{s/\deg f},$$

where the sum is over all monic irreducible polynomials f of F[x] with degree a divisor of s.

Proof. From the theory of finite fields we know that the field F_s with q^s elements is a normal extension of F which splits completely the separable polynomial $X^{q^s} - X$. Therefore any intermediary field between F_s and F is the splitting field of a monic irreducible polynomial which divides $X^{q^s} - x$, whose degree must also divide s.

Let $f \in F[x]$ be a monic irreducible polynomial of degree ds. Let $\alpha_1, \ldots,$ $\alpha_d \in F_s$ be its roots. By Lemma 1 we have $\chi'(\alpha_i)\psi'(\alpha_i) = \Lambda(f)^{s/d}$. Hence $\sum_{i=1}^d \chi'(\alpha_i)\psi'(\alpha_i) = d\Lambda(f)^{s/d}$. Thus summing on the right-hand side over all monic polynomials f of degree a divisor of s is the same as summing $\chi'(\alpha)\psi'(\alpha)$ over all elements α in F_s, i.e. over all roots of $X^{q^s} - X$. This proves Lemma 4.2.

Proof of the Hasse–Davenport relation. Taking the logarithmic derivative of both sides of the Euler product identity

$$1 + G(\chi, \psi) = \prod_f (1 - \Lambda(f)T^{\deg f})^{-1}$$

and multiplying the result by T we obtain

$$\frac{G(\chi, \psi)T}{1 + G(\chi, \psi)T} = \sum_f \frac{(\deg f)\Lambda(f)T^{\deg f}}{1 - \Lambda(f)T^{\deg f}},$$

where the sum on the right-hand side is over all monic irreducible polynomials in $F[x]$. Expressing the denominators in a geometric series yields

$$\sum_{s=1}^\infty (-1)^s G(\chi, \psi)^s T^s = \sum_{r=1}^\infty \sum_f (\deg f)\Lambda(f)^r T^{r \deg f}.$$

Letting $s = r \deg f$, the right-hand side becomes

$$\sum_{s=1}^\infty \left(\sum_{\deg f | s} (\deg f)\Lambda(f)^{s/\deg f} \right) T^s.$$

Comparing the coefficients of equal powers T^s on both sides gives

$$(-1)^{s-1} G(\chi, \psi) = \sum_{\deg f | s} (\deg f)\Lambda(f)^{s/\deg f}.$$

The theorem now follows from the definition of $G(\chi', \psi')$ in Lemma 4.2.

4.3.3 Stickelberger's theorem

Stickelberger's theorem expresses a deep and fundamental property of the Gauss sums. Among other things, it provides a factorization of the principal ideal $(G(\chi, \psi))$ in terms of data related to χ and ψ. The proof is subdivided in three parts; Part II contains the main body of the argument and a precise statement of the theorem; in Part III we give some applications.

Part I   Some combinatorial lemmas

Lemma 4.3 *Let $\varphi(x_1, \ldots, x_f) \in \mathbb{Z}[x_1, \ldots, x_f]$ be invariant under cyclic permutations of the variables:*

$$\varphi(x_2, x_3, \ldots, x_1) = \varphi(x_1, x_2, \ldots, x_f).$$

Let h be a non-negative integer and let ξ be an element in $\mathcal{O} = \mathbb{Z}[\zeta_m]$. Let $\mathcal{Y}$ be a prime ideal in $\mathcal{O}$ of degree f. Then there is a positive integer a such that

$$\varphi(\xi^{p^h}, \xi^{p^{h+1}}, \ldots, \xi^{p^{h+f-1}}) \equiv a \bmod \mathcal{Y}^{h+1}.$$

Proof. We use induction on the integer h. We observe that the multinomial theorem yields the identity

$$\varphi(x_1, \ldots, x_f)^p = \varphi(x_1^p, \ldots, x_f^p) + p\psi(x_1, \ldots, x_f), \qquad (4.1)$$

where ψ is a function of the same kind as φ. First we consider the case $h = 0$: if $\xi \equiv \bar{\xi} \bmod \mathcal{Y}$ and $x = \bar{\xi}$, then the identity (4.1) shows that $\bar{\varphi}(\xi, \xi^p, \ldots, \xi^{p^{f-1}}) = \varphi(x, x^p, \ldots, x^{p^{f-1}})$ is invariant under the Frobenius substitution $y \mapsto y^p$, which generates the Galois group $\mathrm{Gal}(\mathbb{F}_q/\mathbb{F})$ of the residue class field of $\mathcal{Y}$; hence $\varphi(x, x^p, \ldots, x^{p^{f-1}})$ must be an element in $\mathbb{F}_p$ and therefore

$$\varphi(\xi, \xi^p, \ldots, \xi^{p^{f-1}}) \equiv a \bmod \mathcal{Y}$$

for some integer a. Suppose the claim holds for the integer $h - 1$. Then with

$$x_1 = \xi^{p^{h-1}}, \qquad x_2 = \xi^{p^h}, \ldots, x_f = \xi^{p^{f+h-2}}$$

the values of φ and ψ are both congruent $\bmod \mathcal{Y}^h$ to positive integers. Therefore φ^p and $p\psi$ are congruent $\bmod \mathcal{Y}^{h+1}$ to positive integers. The claim now follows from the identity (4.1) for the integer h.

If k is a positive integer, define $\binom{t}{k}$ to be the polynomial $\dfrac{t(t-1)\ldots(t-k+1)}{k!}$; if $k = 0$ put $\binom{t}{k} = 1$. For an integer $a \geq k$, $\binom{a}{k}$ denotes the usual binomial coefficient $\dfrac{a!}{k!(a-k)!}$; if $a < k$ is a positive integer then $\binom{a}{k} = 0$. Therefore, if a is an indeterminate, we have the formal Maclaurin expansion,

$$(1 + t)^a = \sum_{k=0}^{\infty} \binom{a}{k} t^k.$$

Lemma 4.4. *We have*

$$\binom{x_0 + \cdots + x_{f-1}}{k} = \sum_a \binom{x_0}{u_0} \cdots \binom{x_{f-1}}{u_{f-1}},$$

where the sum is over all f-tuples of positive integers $u = (u_0, \ldots, u_{f-1})$ such that $u_0 + \cdots + u_{f-1} = k$.

Proof. This is obvious from the identity

$$(1 + t)^{x_0 + \cdots + x_{f-1}} = \prod_{i=0}^{f-1} (1 + t)^{x_i}$$

and the definition of the polynomials $\binom{a}{k}$.

Lemma 4.5 *There is a section*

$$\mathbb{F}_q^\times = (\mathcal{O}/\mathcal{Y})^\times \to \mathcal{O}, \qquad x \to \mu_x$$

which satisfies $\mu_x^q \equiv \mu_x \bmod \mathcal{Y}^{f+1}$. Further, μ_x is uniquely determined mod $\mathcal{Y}^{f+1}$.

Proof. Let $\mu \in \mathcal{O}$ be such that $\bar{\mu} = x$. Now any element $x \in \mathbb{F}_q$ satisfies $x^q = x$ and hence $\mu^q \equiv \mu \bmod \mathcal{Y}$. Let $\mu_x = \mu^q$. Then $\mu_x \equiv \mu \bmod \mathcal{Y}$ and so $\bar{\mu}_x \equiv \bar{\mu} \equiv x \bmod \mathcal{Y}$. From $\mu_x \equiv \mu \bmod \mathcal{Y}$ we obtain $\mu_x^{p'} \equiv \mu^{p'} \bmod \mathcal{Y}^{p'+1}$ which proves that $\mu_x^q \equiv \mu_x \bmod \mathcal{Y}^{f+1}$. To verify uniqueness let $\mu_x' \in \mathcal{O}$ be another element satisfying $\mu_x'^q \equiv \mu_x \bmod \mathcal{Y}^{f+1}$. Then since $\mu_x' \equiv \mu_x \bmod \mathcal{Y}$, we get $\mu_x'^q \equiv \mu_x' \equiv \mu_x^q \equiv \mu_x \bmod \mathcal{Y}^{f+1}$.

Remark The set

$$M = \{\mu_x : x \in \mathbb{F}_q^\times\}$$

provides a system of representatives in $\mathcal{O}$ for the multiplicative group $\mathbb{F}_q^\times$; in fact, since for $x, y \in \mathbb{F}_q^\times$, the product $\mu_x \mu_y$ satisfies $\overline{\mu_x \mu_y} = xy$ and $(\mu_x \mu_y)^q \equiv \mu_x \mu_y \bmod \mathcal{Y}^{f+1}$ and hence by the uniqueness part of Lemma 4.5 we have

$$\mu_x \mu_y \equiv \mu_{xy} \bmod \mathcal{Y}^{f+1}.$$

The following result captures the orthogonality properties of the characters of the multiplicative group $\mathbb{F}_q^\times$ in terms of the set M.

Lemma 4.6 *Let s be an integer. Then, modulo $\mathscr{Y}^{f+1}$, we have*

$$\sum_{x \in \mathbb{F}_q^{\times}} \mu_x^s \equiv \begin{cases} q - 1 & \text{if } s \in (q-1)\mathbb{Z} \\ 0 & \text{if } s \notin (q-1)\mathbb{Z}. \end{cases}$$

Proof. (i) Suppose $s \in (q-1)\mathbb{Z}$. Then for $x \in \mathbb{F}_q^{\times}$ we have $\mu_x \notin \mathscr{Y}$ and $\mu_x^q \equiv \mu_x \bmod \mathscr{Y}^{f+1}$ implies that $\mu_x^{q-1} \equiv 1 \bmod \mathscr{Y}^{f+1}$. Hence $\mu_x^s \equiv 1 \bmod \mathscr{Y}^{f+1}$ and the sum is $\equiv (q-1)$.

(ii) Suppose $s \notin (q-1)\mathbb{Z}$; by the above remark, we have for any $y \in \mathbb{F}_q^{\times}$

$$\mu_y^s \sum_{x \in \mathbb{F}_q^{\times}} \mu_x^s = \sum_{x \in \mathbb{F}_q^{\times}} (\mu_y \mu_x)^s \equiv \sum_{x \in \mathbb{F}_q^{\times}} \mu_{yx}^s \bmod \mathscr{Y}^{f+1}.$$

Hence $(\mu_y^s - 1) \sum_x \mu_x^s \equiv 0 \bmod \mathscr{Y}^{f+1}$. Since s is not a multiple of $q - 1$ which is the order of the cyclic group $\mathbb{F}_q^{\times}$, there is a choice of element $y \in \mathbb{F}_q^{\times}$, which satisfies $\mu_y^s - 1 \not\equiv 0 \bmod \mathscr{Y}$. This implies

$$\sum_{x \in \mathbb{F}_q^{\times}} \mu_x^s \equiv 0 \bmod \mathscr{Y}^{f+1}.$$

Part II Proof of Stickelberger's theorem

In the following we consider a prime ideal $\mathscr{Y}$ of $\mathbb{Q}(\zeta_m)$ with residue class field $\mathbb{F}_q$, $q = p^f$; we assume that f is the exact order of $p \bmod m$; for otherwise the relevant Gauss sums can be reduced to this case by the Hasse–Davenport relation. We consider the Gauss sum

$$G(\chi, \psi) = \sum_{x \in \mathbb{F}_q^{\times}} \chi(x)\psi(x),$$

where $\chi(x)$ is the multiplicative character defined by the power residue symbol $(x/\mathscr{Y})$ and

$$\psi(x) = \zeta^{\mathrm{Tr}(x)},$$

where $\zeta = \zeta_p = \exp(2\pi i/p)$ and $\mathrm{Tr}(x) = x + x^p + \cdots + x^{p^{f-1}}$. Also observe that a complete set of characters of order m of the multiplicative group $\mathbb{F}_q^{\times}$ is given by χ^a, $1 \le a \le m$, $(a, m) = 1$.

We now combine the basic properties of cyclotomic fields with the lemmas above to obtain a factorization of $G(\chi^a, \psi)$ in $\mathbb{Q}(\mu_{mp})$. As before we fix a prime ideal $\mathscr{F}$ in $\mathbb{Q}(\mu_{mp})$ with $\mathscr{Y} = \mathscr{F} \cap \mathcal{O}$ and let $\mathscr{Y}' = \mathscr{F} \cap \mathbb{Z}[\zeta_p]$. We also have $\mathscr{Y}' = (\pi)$, where $\pi = \zeta - 1$. For any integer a we let

$$g_a = G(\chi^a, \psi).$$

Clearly g_a depends only on the congruence class of $a \bmod m$. In particular $g_a = -1$ if $a \equiv 0 \bmod m$ and $|g_a|^2 = q$ if $a \not\equiv 0 \bmod m$.

We now fix an integer a in the range $0 < a < m$. Since $0 < (m - a)n < mn = p^f - 1$, the p-adic expansion

$$(m - a)n = a_0 + a_1 p + \cdots + a_{f-1} p^{f-1},$$

is valid with each a_i satisfying $0 \le a_i \le p - 1$ and at least one $a_i > 0$ and one $a_i < p - 1$. Let

$$t = a_0 + a_2 + \cdots + a_{f-1}.$$

Then

$$0 < t < (p - 1)f < (p - 1)(f + 1).$$

The essential step in the proof of Stickelberger's theorem is the determination of $g_a \bmod \mathscr{Y}^{f+1}$ by first computing $\chi(x)$ and $\psi(x) \bmod \mathscr{Y}^{f+1}$. We consider first the additive character. By Lemma 4.3, applied to the polynomial $\varphi = x_1 + \cdots + x_f$ with $h = f$ and $\mu = \mu_x$, there is some positive integer a_x such that

$$a_x \equiv \mu_x^{p^f} + \mu_x^{p^{f+1}} + \cdots + \mu_x^{f+f-1} \bmod \mathscr{Y}^{f+1}$$

$$\equiv \mu_x + \mu_x^p + \cdots + \mu_x^{p^{f-1}} \bmod \mathscr{Y}^{f+1}.$$

Therefore

$$\bar{a}_x = x + x^p + \cdots + x^{p^{f-1}} = \sum_\sigma x^\sigma = T(x),$$

where the sum $\sum_\sigma$ runs over all $\sigma \in \mathrm{Gal}(\mathbb{F}_q/\mathbb{F}_p)$. Hence

$$\psi(x) = \zeta^{T(x)} = \zeta^{a_x} = (1 + \pi)^{a_x} = \sum_{k=0}^{\infty} \binom{a_x}{k} \pi^k$$

$$\equiv \sum_{k=0}^{t} \binom{a_x}{k} \pi^k \bmod \mathscr{F}^{t+1},$$

since $\mathscr{F} | \mathscr{Y}$, i.e. $\pi^k \in \mathscr{F}^k$. An elementary calculation shows that if v denotes the valuation of $\mathbb{Q}(\mu_p)$ which uniquely extends the p-adic exponential valuation v of $\mathbb{Q}$ with $v(p) = 1$, and if k is a non-negative integer with p-adic expansion

$$k = c_0 + c_1 p + \cdots + c_r p^r, \qquad 0 \le c_i \le p - 1,$$

then $v(k!) = (k - \sum_{i=0}^{r} c_i)/(p - 1)$. Hence in $\mathbb{Q}(\mu_p)$ $v(\pi^k/k!) = \sum_{i=0}^{r} c_i/(p-1) \ge 0$ and $\pi^k/k!$ is $\mathscr{Y}$-integral; therefore it is also $\mathscr{F}$-integral in $\mathbb{Q}(\mu_{mp})$.

Let $\mu = \mu_x + \mu_x^p + \cdots + \mu_x^{p^{f-1}}$. We know that $a_x \equiv \mu \bmod \mathscr{Y}^{f+1}$. Using the fact that $t < (p - 1)(f + 1)$ and $\mathscr{Y} = \mathscr{F}^{p-1}$ we obtain that $\mathscr{Y}^{f+1} = \mathscr{F}^{(p-1)(f+1)} \subseteq \mathscr{F}^t$ and hence $a_x \equiv \mu \bmod \mathscr{F}^t$. Since clearly

$$\binom{a_x}{k} = a_x(a_x - 1)\ldots(a_x - k + 1)/k! \qquad \text{and} \qquad \pi^k/k!$$

are both $\mathscr{F}$-integral, it follows that

$$\binom{a_x}{k}\pi^k \equiv \binom{\mu}{k}\pi^k \bmod \mathscr{F}^{t+1}.$$

Lemma 4.4 provides the binomial identity

$$\binom{\mu}{k} = \sum_u \binom{\mu_x}{u_0}\cdots\binom{\mu_x^{p^{f-1}}}{u_{f-1}},$$

where the sum $\sum_u$ is taken over all f-tuples $u = (u_0, \ldots, u_{f-1})$ with $u_i \in \mathbb{Z}$, $u_i \geq 0$ and $k = u_0 + \cdots + u_{f-1}$. Let $u = (u_0, \ldots, u_{f-1})$ with $u_i \in \mathbb{Z}$ and $u_i \geq 0$. Then

$$\{x_0(x_0 - 1)\ldots(x_0 - u_0 + 1)\} \cdot \{x_1(x_1 - 1)\ldots(x_1 - u_1 + 1)\}$$
$$\cdots \{x_{f-1}(x_{f-1} - 1)\ldots(x_{f-1} - u_{f-1} + 1)\}$$
$$= \sum_v c_{uv} x_0^{u_0 - v_0} x_1^{u_1 - v_1} \ldots x_{f-1}^{u_{f-1} - v_{f-1}}$$

where the sum $\sum_v$ is taken over all f-tuples $v = (v_0, \ldots, v_{f-1})$ with $v_i \in \mathbb{Z}$, $0 \leq v_i \leq u_i$ and the c_{uv} are suitable integers; observe that the coefficient of the leading term is $c_{u0} = 1$. Using the notation we can now write

$$\binom{\mu}{k} = \sum_{k=0}^{t} \sum_u \sum_v c_{uv} \mu_x^{(u_0 - v_0) + (u_1 - v_1)p + \cdots + (u_{f-1} - v_{f-1})p^{f-1}}.$$

Thus for the character ψ we have

$$\psi(x) \equiv \sum_{k=0}^{t} \sum_u \sum_v c_{uv} \mu_x^{(u_0 - v_0) + (u_1 - v_1)p + \cdots + (u_{f-1} - v_{f-1})p^{f-1}}$$
$$\times \frac{\pi^{u_0 + u_1 p + \cdots + u_{f-1} p^{f-1}}}{(u_0!)\ldots(u_{f-1}!)} \bmod \mathscr{F}^{t+1}.$$

As for the multiplicative character χ we have

$$\chi(x) = (\mu_x / \mathscr{Y}) \equiv \mu_x^n \bmod \mathscr{Y}$$

and so

$$\chi(x)^{p^f} \equiv \mu_x^{np^f} \bmod \mathscr{Y}^{f+1}.$$

But $\chi(x)^{p^f} = \chi(x)$ and $\mu_x^{p^f} \equiv \mu_x \bmod \mathscr{Y}^{f+1}$. Hence $\chi(x) \equiv \mu_x^n \bmod \mathscr{Y}^{f+1}$. Again, since $\mathscr{Y}^{f+1} \subset \mathscr{F}^{t+1}$ we get

$$\chi(x)^a \equiv \mu_x^{na} \bmod \mathscr{F}^{t+1}.$$

From the equality $mn = p^f - 1$ and the p-adic representation

$$(m - a)n + a_0 + a_1 p + \cdots + a_{f-1} p^{f-1}, \qquad 0 \leq a_i \leq p - 1,$$

it follows that

$$an = mn - (m - a)n = (p - 1 - a_0) + (p - 1 - a_1)p + \cdots$$
$$+ (p - 1 - a_{f-1})p^{f-1};$$

hence

$$\chi(x)^a \equiv \mu_x^{(p-1-a_0)+(p-1-a_1)p+\cdots+(p-1-a_{f-1})p^{f-1}} \bmod \mathscr{F}^{t+1}.$$

Now combining the above results on $\psi(x)$ and $\chi(x)$ we obtain

$$g_a \equiv \sum_x \sum_{k=0}^{t} \sum_u \sum_v c_{uv} \mu_x^{e(u,y)} \frac{\pi^k}{u_0! \ldots u_{f-1}!} \bmod \mathscr{F}^{t+1},$$

where the sum $\sum_x$ is taken over all $x \in \mathbb{F}_q^\times$ and $e(u, v) = t_0 + t_1 p + \cdots + t_{f-1} p^{f-1}$, $t_i = (u_i - v_i + p - 1 - a_i) \geq 0$. The sum $\sum_x$ is evaluated using Lemma 4.6:

$$\sum_x \mu_x^s \equiv \begin{cases} q - 1 & \bmod \mathscr{Y}^{t+1} \quad \text{if}(q - 1)|s \\ 0 & \bmod \mathscr{Y}^{t+1} \quad \text{if}(q - 1) \nmid s. \end{cases}$$

Suppose then that $(q - 1)|e(u, v)$, i.e.

$$t_0 + t_1 p + \cdots + t_{f-1}p^{f-1} \equiv 0 \bmod(p^f - 1).$$

This congruence leads to a system of f linear relations in the t_i:

$$t_0 + t_1 p + \cdots + t_{f-1}p^{f-1} = (p^f - 1)d_0, \qquad d_0 \geq 1,$$
$$t_0 p + t_1 p^2 + \cdots + t_{f-1} = (p^f - 1)d_1, \qquad d_1 \geq 1,$$
$$t_0 p^{f-1} + t_1 + \cdots + t_{f-1}p^{f-2} = (p^f - 1)d_{f-1}, \qquad d_{f-1} \geq 1.$$

The sum of both sides of the above equations is

$$\left(\sum_{i=0}^{f-1} t_i\right)(1 + p + \cdots + p^{f-1}) = (p^f - 1)\sum_{i=0}^{f-1} d_i \geq (p^f - 1)f$$

and therefore

$$\sum_{i=0}^{f-1} t_i \geq (p - 1)f.$$

On the other hand

$$\sum_{i=0}^{f-1} t_i = \sum_i (u_i - v_i + p - 1 - a_i)$$
$$\leq \sum_i u_i + (p - 1)f - \sum_i a_i$$
$$= k + (p - 1)f - t \leq (p - 1)f,$$

since $0 \leq k \leq t$. Hence the above inequalities are all equalities and $v_i = 0$ and $k = t$.

Observe that the determinant of the above system of linear relation in the t_i is non-singular, since modulo p it is $\equiv 1$. From the equality $\sum_{i=0}^{f-1} d_i = f$ and the condition $d_i \geq 1$ it follows that all $d_i = 1$. The relation $t_0 + t_1 p + \cdots + t_{f-1} p^{f-1} = p^f - 1$ then forces all $t_i = p - 1$. Hence $p - 1 = t_i = u_i - v_i + p - 1 - a_i$ and from $v_i = 0$ it follows that $u_i = a_i$. That is to say, the only exponent $e(u, v)$ with $(q - 1)|e(u, v)$ is the one corresponding to $v = (0, \ldots, 0)$ and $u = (a_0, \ldots, a_{f-1})$, with $k = t = a_0 + \cdots + a_{f-1}$. The following theorem is now completely proved.

Theorem 4.2 (Stickelberger) *Let a be an integer, $0 < a < m$, and let $(m - a)n = a_0 + a_1 p + \cdots + a_{f-1} p^{f-1}, 0 \leq a_i \leq p - 1$. Let $t = a_0 + a_1 + \cdots + a_{f-1}$. Then*

$$-g_a \equiv \frac{\pi^t}{a_0! \ldots a_{f-1}!} \bmod \mathscr{F}^{t+1}.$$

Part III Some applications of Strickelberger's theorem

We continue using the notation of the previous two sections.

Corollary *In the cyclotomic field $\mathbb{Q}(\mu_{mp})$, the ideal generated by g_a satisfies $\mathscr{F}^t \| g_a$. Furthermore in $\mathbb{Q}(\mu_m)$ we have*

$$\mathscr{Y}^{tm/(p-1)} \| g_a^m.$$

Proof. Since the a_i satisfy $0 \leq a_i < p$, it is clear that $\mathscr{F}$ does not divide the $a_i!$. Now the ideal $\mathscr{Y}' = (\pi)$ in $\mathbb{Q}(\mu_p)$ does not ramify in $\mathbb{Q}(\mu_{mp})$; hence $\mathscr{F} \| \pi$ and therefore by Stickelberger's theorem we have $\mathscr{F}^t \| g_a$. As for the second assertion we observe that $g_a^m \in \mathbb{Q}(\mu_m)$ and $\mathscr{F}^{p-1} = \mathscr{Y}$; hence $\mathscr{F}^{(p-1)tm/(p-1)} \| g_a^m$. This proves the corollary.

For each positive integer i let r_i be the smallest positive residue of $(m - a)p^i \bmod m$, i.e., $(m - a)p^i \equiv r_i \bmod m$ and $0 \leq r_i < m$. Since $mn = q - 1 = p^f - 1$ and $(m - a)n = a_0 + a_1 p + \cdots + a_{f-1} p^{f-1}$ we obtain

$$nr_i \equiv (m - a)np^i \bmod (p^f - 1)$$

$$\equiv (a_0 p^i + a_1 p^{i+1} + \cdots + a_{f-1} p^{f-1+i}) \bmod (p^f - 1)$$

$$\equiv (a_0 p^i + \cdots + a_{f-1-i} p^{f-1} + a_{f-i} + \cdots + a_{f-1} p^{i-1}) \bmod (p^f - 1).$$

Now both sides are positive and, since not all a_i are equal to $p - 1$, both are $< p^f - 1$; therefore the two sides are equal:

$$nr_i = a_0 p^i + \cdots + a_{f-1-i} p^{f-1} + a_{f-i} + \cdots + a_{f-1} p^{i-1};$$

Hence

$$n \sum_{i=0}^{f-1} r_i = \left(\sum_{i=0}^{f-1} a_i \right) (1 + p + \cdots + p^{f-1}) = tmn/(p-1),$$

or

$$\sum_{i=0}^{f-1} r_i = tm/(p-1).$$

As usual for any real number α we denote by $[\alpha]$ the greatest integer $\leq \alpha$ and by $\langle \alpha \rangle = \alpha - [\alpha]$, the fractional part of α. Using this notation and the definition of the r_i we obtain

$$r_i/m = \langle (m-a)p^i/m \rangle = \langle -ap^i/m \rangle$$

and

$$tm/(p-1) = \sum_{i=0}^{f-1} r_i = m \sum_{i=0}^{f-1} \langle -ap^i/m \rangle.$$

Hence the second assertion of the corollary is equivalent to

$$\mathscr{Y}^{m \sum_{i=0}^{f-1} \langle -ap^i/m \rangle} \| g_a^m.$$

Clearly this last statement depends only on the residue class of $a \bmod m$. Thus we may remove the restriction that $0 < a < m$ and in the following we write

$$\mathscr{Y}^{m \sum_{i=0}^{f-1} \langle -ap^i/m \rangle} \| g_a^m \qquad \text{for all } a \not\equiv 0 \bmod m.$$

Under the isomorphism $(\mathbb{Z}/m\mathbb{Z})^\times \overset{\sim}{\to} \mathrm{Gal}(\mathbb{Q}(\mu_m)/\mathbb{Q})$, the images of the f distinct residue classes $\{ \bar{1}, \bar{p}, \bar{p}^2, \ldots, \bar{p}^{f-1} \}$ correspond to the decomposition group D of the ideal $\mathscr{Y}$ in $\mathbb{Q}(\mu_m)$:

$$D = \{ \sigma \in \mathrm{Gal}(\mathbb{Q}(\mu_m)/\mathbb{Q}) : \mathscr{Y}^\sigma = \mathscr{Y} \}.$$

Let u be an integer relatively prime to m and observe that the automorphism $\sigma_u : \zeta_m \to \zeta_m^u$ depends only on the residue class $\bar{u}$ of $u \bmod m$; hence

$$D = \{ \sigma_{p^i} : 0 \leq i \leq f-1 \}.$$

Let

$$\mathrm{Gal}(\mathbb{Q}(\mu_m)/\mathbb{Q}) = \bigcup_1^g \sigma_{u_i}^{-1} D$$

be a coset decomposition of the Galois group with respect to D, where the u_i are integers relatively prime to m. From the theory of cyclotomic fields we know that in $\mathbb{Q}(\mu_m)$ the ideal (p) has a factorization into distinct factors given by

$$(p) = \mathscr{Y}^{\sigma_{u_1}^{-1}} \dots \mathscr{Y}^{\sigma_{u_g}^{-1}}.$$

Since each σ_{p^i} leaves $\mathscr{Y}$ fixed we can write

$$\mathscr{Y}^{m\sum_{i=0}^{f-1}\langle -ap^i/m\rangle} = \mathscr{Y}^{m\sum_{i=0}^{f-1}\langle -ap^i/m\rangle\sigma_{p^i}^{-1}}.$$

From the properties of the Gauss sums we have for any integer u prime to m

$$\sigma_u(g_a^m) = g_{ua}^m.$$

Hence if $\mathscr{Y}^E \| g_{ua}^m$, i.e. $\mathscr{Y}^E \| (\sigma_u g_a)^m$, then

$$\mathscr{Y}^{\sigma_u^{-1}E} \| g_a^m,$$

and the $\mathscr{Y}^{\sigma_{u_i}^{-1}}$ component of (g_a^m) is

$$(\mathscr{Y}^{m\sum_{i=0}^{f-1}\langle -au_jp^i\rangle\sigma_{p^i}^{-1}})^{\sigma_{u_j}^{-1}} = \mathscr{Y}^{m\sum_{i=0}^{f-1}\langle -au_jp^i\rangle\sigma_{p^i}^{-1}u_j}.$$

From the factorization $q = g_a \cdot \bar{g}_a$ it follows that the only prime ideals which divide g_a^m are those which divide (p), i.e. the $\mathscr{Y}^{\sigma_{u_j}^{-1}}$. Hence

$$(g_a^m) = \mathscr{Y}^{m\sum_r \langle -ra/m\rangle\sigma_r^{-1}}$$

where the sum is over a set of representative residues of $(\mathbb{Z}/m\mathbb{Z})^\times \simeq \mathrm{Gal}(\mathbb{Q}(\mu_m)/\mathbb{Q})$. We thus have

Theorem 4.3 (Stickelberger) *Let $a \in \mathbb{Z}$, $(a, m) = 1$ and let*

$$\Phi(a) = \sum_{\substack{0 < r < m \\ (r,m)=1}} \langle -ar/m\rangle\sigma_r^{-1}.$$

Then

$$(g_a^m) = \mathscr{Y}^{\Phi(a)}.$$

Remark The last theorem is a cornerstone in the theory of abelian extensions of cyclotomic fields.

Definition of Jacobi sums 4.1 *Let $a, b \in \mathbb{Z}$ and put*

$$J_{a,b} = \sum_x{}' \mu(x)^a\mu(1-x)^b,$$

where the sum runs over $x \in \mathbb{F}_q^\times - \{1\}$ and μ is a multiplicative character. It is clear that these sums are elements in $\mathbb{Q}(\mu_m)$.

Theorem 4.4 *Let $a, b \in \mathbb{Z}$ be prime to m with $m \nmid a + b$. Then*

$$g_a g_b = g_{a+b} J_{a,b}.$$

In particular $J_{a,b} \neq 0$.

Proof. By defining $\mu(0) = 0$, we have

$$g_a g_b = \sum_x \psi(x)\mu(x)^a \sum_y \psi(y)\mu(y)^b, \qquad x, y \in \mathbb{F}_q^\times.$$

$$= \sum_{x,y} \psi(x + y)\mu(x)^a \mu(y)^b$$

$$= \sum_{y,z} \psi(z)\mu(z - y)^a \mu(y)^b, \qquad z = x + y$$

where the sum now runs over all $z \in \mathbb{F}_q$ and $y \in \mathbb{F}_q^\times$. To calculate the sum over y we observe that if $z = 0$ then

$$\sum_y \mu(z - y)^a \mu(y)^b = (-1)^a \sum_y \mu(y)^{a+b} = 0,$$

since $m \nmid a + b$. If $z \neq 0$, put $y = zu$ and observe that the sum becomes

$$= \mu(z)^{a+b} \sum_u \mu(1 - u)^a \mu(u)^b.$$

Thus

$$g_a g_b = \sum_z \mu^{a+b}(z) \sum_u \mu(1 - u)^a \mu(u)^b$$

$$= g_{a+b} \cdot J_{a,b}.$$

This proves the theorem.

4.4 Kloosterman sums

4.4.1 Second example of an *L*-function for the projective line

Kloosterman sums arise in many number theoretic problems, e.g. in the estimation of the Fourier coefficients of modular forms. They also play a role in the study of the representations of the group GL_2 over finite fields and p-adic fields; in this group-theoretic context they are the analogue of the K-Bessel functions with which they have many properties in common. In this introductory section we consider some elementary properties of Kloosterman sums related to *L*-functions on the projective line.

Consider the group $G = \mathbb{F}_q \times \mathbb{F}_q$ and let $\hat{G}$ be its character group. For χ an element in $\hat{G}$ we write $\chi = \psi\varphi$, where $\psi = \mathrm{Res}(\chi)_{\mathbb{F}_q \times \{0\}}$ and $\varphi = \mathrm{Res}(\chi)_{\{0\} \times \mathbb{F}_q}$, i.e. the product of its restrictions to the first and second components.

Definition 4.1 Let $\chi \in \hat{G}$ have the decomposition $\chi = \psi\varphi$. The Kloosterman sum associated to χ is

$$K(\chi) = - \sum_{x \in F_q^\times} \psi(x)\varphi(x^{-1})$$

More generally, if we define for each positive integer n, $G_n = F_{q^n} \times F_{q^n}$ and consider the corresponding embedding $\hat{G}_1 \hookrightarrow \hat{G}_n$ given by

$$\chi_1 \mapsto \chi_1 \circ \mathrm{Tr} = \chi_n,$$

where $\mathrm{Tr}(a, b) = (\mathrm{Tr}_{F_n/F_1}(a), \mathrm{Tr}_{F_n/F_1}(b))$, $F_n = F_{q^n}$, then it is possible to define a Kloosterman sum by the formula

$$K(\chi_n) = -\sum_P \chi_n(P).$$

where the sum $\sum_P$ runs over all the points in the hyperbola

$$H(F_n) = \{(x, x') \in F_n \times F_n : xx' = 1\}$$

We show later, using a theorem of Carlitz, that the sums $K(\chi_n)$ satisfy a Hasse–Davenport type relation.

Consider the family of polynomials in T which do not vanish at the origin on the affine line A^1

$$A = \{a = T^m + a_1 T^{m-1} + \cdots + a_m \in F_q[T], a_m \neq 0\},$$

we assume that $1 \in A$. Let $\chi \in \hat{G}$ have the decomposition $\chi = \psi\varphi$. For an element $a = T^m + a_1 + \cdots + a_m$, put

$$\Lambda(a) = \psi(a_1)\varphi(a_{m-1}\bar{a}_m), \qquad \bar{a}_m a_m = 1;$$

also let $\Lambda(1) = 1$. As in Section 4.2, if $a = T^m + a_1 T^{m-1} + \cdots + a_{m-1} T + a_m$, $b = T^n + b_1 T^{n-1} + \cdots + b_{n-1} T + b_n$, and

$$a \cdot b = T^{m+n} + (a_1 + b_1)T^{m+n-1} + \cdots + (a_{m-1}b_n + b_{n-1}a_m)T + a_m b_n;$$

hence

$$\begin{aligned}
\Lambda(a \cdot b) &= \psi(a_1 + b_1)\varphi((a_{m-1}b_n + b_{n-1}a_m)\bar{a}_m \cdot \bar{b}_n) \\
&= \psi(a_1)\psi(b_1)\varphi(a_{m-1}\bar{a}_m + b_{n-1}\bar{b}_n) \\
&= \Lambda(a)\Lambda(b).
\end{aligned}$$

Thus we can define a new L-function by putting

$$Z(s, K, P^1) = \sum_a \Lambda(a)Na^{-s}$$

$$= \prod_P \frac{1}{1 - \Lambda(P)NP^{-s}},$$

where the sum is extended over the set of polynomials $a \in A$, including the polynomial $a = 1$, and the product is taken over the subset of irreducible polynomials in A.

By grouping the terms in the Dirichlet series $Z(s, K, P^1)$ corresponding to polynomials of the same degree we obtain

$$Z(s, K, P^1) = 1 + \sum_{d=1}^{\infty} q^{-ds} S_d,$$

where

$$S_d = \sum_a \Lambda(a),$$

and the sum $\sum_a$ is extended over all monic polynomials $a \in A$ of degree d. The values of S_d are completely determined below.

$(d = 1)$: $\Lambda(T + a) = \psi(a_1)\varphi(\bar{a}_1)$;

$(d = 2)$: $\Lambda(T^2 + a_1 T + a_2) = \psi(a_1)\varphi(a_1 \bar{a}_2)$;

$(d > 2)$:

$$\sum_{a \in A,\, \deg(a)=d} \Lambda(a) = \sum_{a_1,\ldots,a_{d-1} \in F_q,\, a_d \in F_q^{\times}} \Lambda(T^d + a_1 T + \cdots + a_{d-1} T + a_d),$$

$$= q^{d-s} \sum_{a_1 \in F_q} \psi(a_1) \sum_{a_{d-1} \in F_q,\, a_d \in F_q^{\times}} \varphi(a_{d-1} \bar{a}_d)$$

$$= q^{m-1}(q - 1)\delta_\Lambda;$$

where $\delta_\Lambda = 1$ if Λ is the trivial character and 0 otherwise.

If the restriction of χ to one of the factor $F_q \times F_q$ is trivial but not the other, then by the orthogonality relations we have

$$\sum_{a \in A,\, \deg(a)=2} \Lambda(T^2 + a_1 T + a_2) = \sum_{a_2 \in F_q^{\times}} \sum_{a_1 \in F_q} \psi(a_1)\varphi(a_1 \bar{a}_2) = 0.$$

If $\chi = \psi\varphi$ is generic, i.e. ψ and φ are both non-trivial, then

$$\sum_{a \in A,\, \deg(a)=2} \Lambda(T^2 + a_1 T + a_2) = \sum_{a_2 \in F_q^{\times}} \sum_{a_1 \in F_q} \psi(a_1)\phi(a_1 \bar{a}_2)$$

$$= \sum_{a_2 \in F_q^{\times}} \sum_{a_1 \in F_q} \varphi_b(a),$$

where $\varphi_b(a) = \psi(a)\varphi(a\bar{b})$ is a character of F_q. Since $\psi\varphi$ is generic, $\psi(x) = \varphi(cx)$ for some $c \in F_q^{\times}$; hence $\psi(a)\varphi(ab) = \varphi(ca)\varphi(ab) = \varphi(a(c + b))$. Therefore the sum $\sum_a \varphi_b(a)$ is 0 if $c + b \in F_q^{\times}$; since $\psi\varphi = 1$ if $c + b = 0$, we see that the whole sum $\sum_a \Lambda(a)$ is q. We also have that if the factor ψ is not trivial but the other is, then

$$\sum_{a \in A,\, \deg(a)=1} \Lambda(a) = \sum_{a_1 \in F_q^{\times}} \psi(a_1) = -1.$$

If χ is generic, i.e. both ψ and φ are non-trivial, then

$$\sum_{a \in A, \deg(a)=1} \Lambda(a) = \sum_{a \in \mathbb{F}_q} \psi(a)\varphi(\bar{a}) = -K(\chi),$$

the ordinary Kloosterman sum.

In the above observations we have proved the following lemma.

Lemma 4.7 *With notations as above, the sum*

$$S_d(\Lambda) = \sum_{a \in A, \deg(a)=d} \Lambda(a)$$

satisfies:

(i) $S_1(\Lambda) = -K(\chi)$, *if both components of χ are non-trivial.*
(ii) $S_1(\Lambda) = -1$, *if one but not the other component of χ is trivial.*
(iii) $S_d(\Lambda) = q^{d-1}(q-1)$, *if $\chi \equiv 1$ (the trivial character). ($d \geq 1$.)*
(iv) $S_d(\Lambda) = 0$, *if $d \geq 3$ and χ is non-trivial.*
(v) $S_2(\Lambda) = q$, *if both components of χ are non-trivial.*
(vi) $S_2(\Lambda) = 0$, *if one but not the other component of χ is trivial.*

From the representation $Z(s, K, P^1) = 1 + \sum_d q^{-ds} S_d(\Lambda)$, we now obtain that

(i) $Z(s, K, P^1) = 1 - K(\Lambda)q^{-s} + q^{1-2s}$ *if both components of χ are nontrivial.*
(ii) $Z(s, K, P^1) = 1 - q^{-s}$ *if only one component of χ is trivial.*
(iii) $Z(s, K, P^1) = (1 - q^{-s})(1 - q^{1-s})^{-1}$ *if χ is trivial.*

As a concrete example we recall that if $k = \mathbb{F}_p$ and $\psi_k(x) = e^{2\pi i x/p}$, then with the choice of character $\chi(x) = \psi(x)\varphi(x)$ with $\psi(x) = \psi_k(ax)$ and $\varphi(x) = \psi_k(bx)$, we obtain the usual Kloosterman sum

$$S_1(\chi) = \sum_{x \in \mathbb{F}_q^\times} \Lambda(T - x)$$

$$= \sum_{x \in \mathbb{F}_q^\times} \psi_k(ax + bx^{-1}).$$

4.4.2 A Hasse–Davenport relation for Kloosterman sums

Theorem 4.5 (Carlitz) *For any character χ_1 of $G_1 = \mathbb{F}_q \times \mathbb{F}_q$ we have*

$$-\log(1 - K(\chi_1)q^{-s} + q^{1-2s}) = \sum_{n=1}^{\infty} K(\chi_n)\frac{q^{-sn}}{n}.$$

Proof. By definition we have, with $\chi_1 = \psi\varphi$,

$$K(\chi_n) = -\sum_{x \in \mathbb{F}_{q^n}^\times} \psi(\mathrm{Tr}_n(x))\varphi(\mathrm{Tr}_n(\bar{x})).$$

where $\mathrm{Tr}_n(x) = x + x^q + \cdots + x^{q^{n-1}}$. This can also be expressed as

$$K(\chi_n) = \sum_{u,v \in F_q} \psi(u)\varphi(v)h_n(u,v)$$

$$= -\sum_g \chi(g)h_n(g),$$

where the sum $\sum_g$ runs over all elements $g \in G_1$, and

$$h_n(u,v) = \sum_{x \in F_{q^n}^\times} 1$$

and $\sum_x$ counts those x with $\mathrm{Tr}_n(x) = u$ and $\mathrm{Tr}_n(\bar{x}) = v$.

From the Euler product representation

$$Z(s, K, P^1) = \prod_P \frac{1}{1 - \Lambda(P)NP^{-s}},$$

where $\prod_P$ is extended over all monic irreducible polynomials in A excluding $P = T$ and $P = 1$. Taking logarithms of both sides we obtain

$$\log Z(s, K, P^1) = \sum_P \sum_{n=1}^{\infty} \Lambda(P^k)(NP^k)^{-s}/k$$

$$= \sum_{m=1}^{\infty} (q^{-ms}/m) \sum_{k=1, k|m} \frac{m}{k} \sum_{P, \deg P = m/k} \Lambda(P^k).$$

Now, if P is a monic irreducible polynomial of degree $r = m/k$, then it has a factorization

$$P(T) = (T - \xi)(T - \xi^q)\ldots(T - \xi^{q^{r-1}}).$$

where ξ is a primitive element of F_{q^n}, that is to say any element of F_{q^n} is some power of ξ. Let

$$P(T) = T^r + a_1 T^{r-1} + \cdots + a_r, \ a_i \in F_q, \ a_r \neq 0.$$

Comparing coefficients and using some properties of the elementary symmetric functions we get

$$-a_1 = \xi + \xi^q + \cdots + \xi^{q^{r-1}}$$

$$-a_{r-1}\bar{a}_r = \bar{\xi} + \bar{\xi}^q + \bar{\xi}^{q^{r-1}}, \ \xi\bar{\xi} = 1.$$

On the other hand the binomial theorem gives

$$P(T)^k = T^{kr} + ka_1 T^{kr-1} + \cdots + ka_{r-1}a_r^{k-1}T + a_r^k.$$

Now

$$\mathrm{Tr}_{kr}(\xi) = \xi + \xi^q + \cdots + \xi^{q^{kr-1}} = k(\xi + \xi^q + \cdots + \xi^{q^{r-1}}) = -ka_1;$$

similarly $\mathrm{Tr}_{kr}(\xi) = -ka_{r-1}\bar{a}_r$. If we put $m = kr$, then

$$\Lambda(P^k) = \psi(\mathrm{Tr}_n(\xi))\varphi(\mathrm{Tr}_n(\xi)) = \chi_n(\xi).$$

To each irreducible polynomial P of degree r there correspond r values of ξ, namely $\xi, \xi^q, \ldots, \xi^{q^{r-1}}$. Thus

$$\sum_{k|n} \frac{n}{k} \sum_{P \in 1, \deg P = n/k} = \sum_{b,c \in F_q} \psi(b)\varphi(c) \sum_{\xi \in F_q^\times, \, \mathrm{Tr}_n(\xi)=n, \, \mathrm{Tr}_n(\bar{\xi})=c} 1$$

$$= \sum_{b,c \in F_q} \psi(b)\varphi(c)h_n(b,c)$$

$$= \sum_{g \in G} \chi(g) \sum_{Q \in H(F_{q^n}), \, Q \to g} 1.$$

Therefore we obtain

$$\log Z(s, K, P^1) = \sum_{n=1}^{\infty} \frac{q^{-ns}}{n} \sum_{g \in G} \chi(g)h_n(g).$$

Finally, if both components of χ are non-trivial we obtain from the Lemma in Section 4.4.1 that

$$-\log(1 - K(\chi)q^{-s} + q^{1-2s}) = \sum_{n=1}^{\infty} q^{-ns}K(\chi_n)/n.$$

This proves the theorem.

4.5 Third example of an *L*-function for the projective line

Let Λ be the set of polynomials in $\mathbb{F}_q[T]$, monic and with non-zero constant term. As before let ψ and φ be two additive characters of $\mathbb{F}_q^+$. Suppose that $\mathbb{F}_q$ is of odd characteristic. For each polynomial

$$P = T^n + a_1 T^{n-1} + \cdots + a_{n-2}T^2 + a_{n-1}T + a_n, \qquad a_n \neq 0,$$

we define a function

$$\Lambda(P) = \psi(a_1)\varphi(a_{n-2}\bar{a}_n - \tfrac{1}{2}a_{n-1}^2\bar{a}_n^2).$$

Lemma 4.8 *We have, with notations as above,*

$$\Lambda(P)\Lambda(Q) = \Lambda(PQ).$$

Proof. If $Q = T^m + b_1 T^{m-1} + \cdots + b_{m-2}T^2 + b_{m-1}T + b_m$, $b_m \neq 0$, we have

$$PQ \equiv (a_{n-2}b_m + a_{n-1}b_{m-1} + a_n b_{m-2})T^2$$
$$+ (b_{m-1}a_n + b_m a_{n-1})T + a_n b_m \bmod T^3.$$

From this we obtain easily that

$$\frac{(a_{n-2}b_m + a_{n-1}b_{m-1} + a_n b_{m-2})}{a_n b_m} - \frac{1}{2}\left(\frac{b_{m-1}a_n + b_m a_{n-1}}{a_n b_m}\right)^2$$

$$= \frac{a_{n-2}}{a_n} - \frac{1}{2}\left(\frac{a_{n-1}}{a_n}\right)^2 + \frac{b_{m-2}}{b_m} - \frac{1}{2}\left(\frac{b_{m-1}}{b_m}\right)^2.$$

This proves the lemma.

As in previous examples we can define a new *L*-function for the projective line by putting

$$Z(s, \Lambda, P^1) = \sum_a \Lambda(a) N a^{-s}$$

$$= \prod_P \frac{1}{1 - \Lambda(P) N P^{-s}},$$

where $\sum_a$ is extended over all monic polynomials with non-zero constant term and $\prod_P$ is extended over all monic irreducible polynomials $P \neq x$. If we let $u = q^{-s}$, then $Z(s, \Lambda, P^1)$ is a polynomial in u. The coefficient of u is the sum

$$\sum_{a \in F_q^\times} \Lambda(T - a).$$

The simplest non-trivial choice for Λ is $\Lambda = \psi \varphi$ with $\psi = \varphi = e^{2\pi i(*)/p}$. In this case one is led to the exponential sum

$$\sum_{a \in F_q^\times} e^{2\pi i\{a - (1/2)a^{-2}\}/p}.$$

In Section 4.6.2 we shall obtain bounds for the above sum as well as for the Kloosterman sum similar to those that are possible for Gauss sums.

4.6 Basic arithmetic theory of exponential sums

4.6.1 Part I: *L*-functions for the projective line

In this section we shall explain the elementary aspects of the connection between exponential sums and characters of the class groups of the projective line. Following Weil, we use these ideas to obtain bounds for exponential sums from the Riemann hypothesis for function fields.

As before, F_q stands for the finite field of q elements and $F_{q^d} = F_d$ is its unique extension of degree d. The rational function field $F_q(T)$, with T a transcendental element, is considered as the field of functions on the projective line P^1/F_q.

As usual, divisors on $\mathbb{P}^1$ are represented by

$$D = \sum_v m_v P_v.$$

When the point P_∞ at infinity appears with multiplicity $m_\infty = 0$, we say that D is a finite divisor. Since the projective line is obtained from the affine line by adding the point at infinity, such divisors can be thought of as divisors on the affine line $\mathbb{A}^1$. In the following we agree to think of $\pi = 1/T$ as the local uniformizing parameter for the point P_∞ at infinity. The finite positive divisors D on $\mathbb{P}^1$ correspond to ideals I in $\mathbb{F}_q[T]$. Suppose that under the correspondence $D \mapsto I$ the polynomial $P_I(T) = T^d + a_1 T^{d-1} + \cdots + a_d$ is the monic polynomial generating I. The degree of D is d. An arbitrary finite divisor can be expressed as the difference of two positive divisors

$$D = D_0 - D_\infty;$$

under the correspondence $D_0 \mapsto I_0$ and $D_\infty \mapsto I_\infty$, we can associate to D the rational function

$$f_D(T) = P_{I_0}(T)/P_{I_\infty}(T).$$

D is the divisor of a function on $\mathbb{P}^1$ if and only if $\deg(D) = 0$, i.e. when $\deg P_{I_0} = \deg P_{I_\infty}$. In this case f_D is the unique function in $\mathbb{F}_q(T)$, with value 1 at P_∞ and satisfying $D = (f_D)$.

Let $D = \sum_v m_v P_v$ be a finite divisor of $\mathbb{P}^1$ and let f be a rational function in $\mathbb{F}_q(T)$ whose divisor (f) has support disjoint from the support of D. The value of f at D is defined by the expression

$$f(D) = \prod_v f(P_v)^{m_v},$$

where for a closed point P of degree d, $f(P) = \mathrm{Norm}_{\mathbb{F}_{q^d}/\mathbb{F}_q}(f(\xi))$ and ξ is an element in $\mathbb{F}_{q^d}$ corresponding to P. Equivalently, if P corresponds to the polynomial $T^d + a_1 T^{d-1} + \cdots + a_d = \prod_{i=1}^d (T - \xi_i)$, $\xi_i \in \mathbb{F}_{q^d}$, then $f(P) = \prod_{i=1}^d f(\xi_i)$.

In the following we consider a character of the multiplicative group

$$\chi: \mathbb{F}_q^\times \mapsto C^\times$$

and fix once and for all a finite positive divisor

$$\mathscr{F}_0 = \sum_v a_v P_v,$$

and suppose that none of the multiplicities a_v are multiples of the order of χ. As the field of formal power series satisfies $\mathbb{F}_q((T)) = \mathbb{F}_q((1/T))$, we can write

$$\mathbb{F}_q((T))^\times \cong \{T^\mathbb{Z}\} \times U,$$

where $\{T^\mathbb{Z}\}$ is the cyclic group generated by T and U is the subgroup of invertible power series in T^{-1}. For each positive integer N we define a subgroup

$$U_N = \{f \in U : f \equiv 1 \bmod T^{-N}\};$$

we clearly have $U \cong \mathbb{F}_q^\times \times U_1$. The family $\{U_N\}_{N \in \mathbb{N}}$ forms a complete system of neighborhoods of the identity and under the induced topology U is a compact group; thus it makes sense to talk about continuous characters

$$\omega \colon \mathbb{F}_q((T))^\times \cong \mathbb{F}_q^\times \times \{T^\mathbb{Z}\} \times U_1 \to \mathbb{C}^\times.$$

We suppose that $\omega(aT) = 1$ for any $a \in \mathbb{F}_q^\times$. By continuity there is some N such that $\omega(U_N) = 1$.

Definition 4.2 *The ideal (T^{-N}) in $\mathbb{F}_q[1/T]$ is called the conductor of ω if $\omega(f) = 1$ for every $f \equiv 1 \bmod T^{-N}$ and N is the smallest integer with this property. In this case we have $\ker \omega = \mathbb{F}_q^\times \times \{T^\mathbb{Z}\} \times U_N$.*

Remark For $i \geq 1$, the map given by

$$x \mapsto (1 + x) \bmod U_i$$

induces an isomorphism

$$(T^{-i})/(T^{-i-1}) \to U_i/U_{i+1}$$

which shows that the quotient

$$\mathbb{F}_q((T))^\times/\ker \omega \cong U_1/U_N$$

is a finite p-group isomorphic (non-canonically) to number of copies of the additive group $\mathbb{F}_q^+$. Thus the values are p^s-roots of 1, $p = \mathrm{char}\, \mathbb{F}_q$.

For each $a \in \mathbb{F}_q$ we consider the function

$$\lambda(a) = \omega\left(\frac{1}{T} - a\right).$$

Given a function $f \in \mathbb{F}_q((T))^\times = \mathbb{F}_q((1/T))^\times$ we expand it in terms of the local uniformizing parameter at infinity and denote by $f(1/T)$ the formal power series in $1/T$; the expression $\omega(f(1/T))$ is clearly well defined. For every finite divisor D whose support is disjoint from $\mathscr{F}_0$ we define a function

$$\Lambda(D) = \omega\left(f_D\left(\frac{1}{T}\right)\right)\chi(f_D(\mathscr{F}_0));$$

$\Lambda(D)$ satisfies

$$\Lambda(D + D') = \Lambda(D)\Lambda(D').$$

Let N be the exponential conductor of ω and put $\mathcal{F} = \mathcal{F}_0 + NP_\infty$. We define two sub-groups of divisors in $\mathrm{Div}(\mathbb{P}^1)$:

$$\mathrm{Div}(\mathcal{F}) = \{D \in \mathrm{Div}(\mathbb{P}^1): \mathrm{supp}\, D \cap \mathcal{F} = \varnothing\}$$

$\mathrm{Div}_0(\mathcal{F})$

$$= \left\{D = (f): f \in \mathbb{F}_q(T)^\times, f\left(\frac{1}{T}\right) \equiv 1 \bmod T_{-N}, f(P_v) = 1 \text{ for all } P_v \in \mathrm{supp}\, \mathcal{F}_0\right\}$$

and put $\mathrm{Cl}_{\mathcal{F}}(\mathbb{P}^1) = \mathrm{Div}(\mathcal{F})/\mathrm{Div}_0(\mathcal{F})$. The function Λ defines a character

$$\Lambda: \mathrm{Cl}_{\mathcal{F}}(\mathbb{P}^1) \to \mathbb{C}^\times.$$

Remark The multiplicities a_v in the divisor $\mathcal{F}_0 = \sum_v a_v P_v$ do not enter into the definition of $\mathrm{Cl}_{\mathcal{F}}(\mathbb{P}^1)$. In fact if χ is a nontrivial character then the proper conductor of Λ is $NP_\infty + \sum_v P_v$. In the following we shall treat Λ as if it were a primitive character.

Definition 4.4 *The L-function of the projective line $\mathbb{P}^1/\mathbb{F}_q$ corresponding to the character $\Lambda: \mathrm{Cl}_{\mathcal{F}}(\mathbb{P}^1) \to \mathbb{C}^\times$ is*

$$L(s, \Lambda, \mathbb{P}^1) = \sum_D \Lambda(D)ND^{-s}$$

$$= \prod_P \frac{1}{1 - \Lambda(P)NP^{-s}},$$

where $\sum_D$ is taken over all finite positive divisors D with $\mathrm{supp}\, D \cap \mathcal{F} = \varnothing$ and $\prod_P$ is taken over all closed point $P \notin \mathrm{supp}(\mathcal{F})$.

Remark $L(s, \Lambda, \mathbb{P}^1)$ is the analogue for the rational function field $\mathbb{F}_q(T)$ of Dirichlet's L-functions.

By class field theory there is an Abelian covering $\mathcal{C} \to \mathbb{P}^1$ of the projective line, i.e. there is a finite separable extension K of $\mathbb{F}_q(T)$ with Galois group isomorphic to $\mathrm{Cl}_{\mathcal{F}}(\mathbb{P}^1)$, having the curve $\mathcal{C}$ as a model. The decomposition law which describes how closed points in $\mathbb{P}^1$ split in the covering $\mathcal{C}$ is equivalent to the following identity between zeta and L-functions

$$Z(s, \mathcal{C}) = Z(s, \mathbb{P}^1) \prod_i L(s, \Lambda_i, \mathbb{P}^1),$$

where $Z(s, \mathcal{C})$ is the zeta function of $\mathcal{C}/\mathbb{F}_q$, $Z(s, \mathbb{P}^1)$ is the zeta function of

$\mathbb{P}^1/\mathbb{F}_q$ and the product $\prod_i$ is taken over all the non-trivial characters of $\text{Gal}(\mathscr{C}/\mathbb{P}^1)$ one of which is the original character Λ. In the above identity, after dividing by $Z(s, \mathbb{P}^1)$, we get a polynomial identity in $u = q^{-s}$; comparing degrees of both sides we get an anlogue of the Hurwitz formula

$$2g(\mathscr{C}) = \sum_i^f (f(\Lambda_i) - 2),$$

where $f(\Lambda_i)$ is the conductor of Λ_i. As the genus of $\mathbb{P}^1$ is 0, by the results of Chapter 3, we have that $L(s, \Lambda, \mathbb{P}^1)$ is a polynomial in $u = q^{-s}$ of degree

$$r = N + d - 2,$$

where d is the number of distinct closed points in the finite divisor $\mathscr{F}_0 = \sum_v a_v P_v$.

From the expression of $L(s, \Lambda, \mathbb{P}^1)$ as a factor of $Z(s, \mathscr{C})$ and the Riemann hypothesis applied to the curve $\mathscr{C}$ we get that the roots of the polynomial

$$L(s, \Lambda, \mathbb{P}^1) = \sum_D \Lambda(D) u^{\deg D}$$

$$= \prod_{i=1}^r (1 - \alpha_i u)$$

satisfy $|\alpha_i| = q^{1/2}$. The coefficient of u corresponds to the number of closed points P defined over $\mathbb{F}_q$; equating coefficients of u on both sides gives

$$\sum_D{}' \Lambda(D) = -\sum_{i=1}^r \alpha_i,$$

where $\sum_D'$ is taken over all finite positive divisors of degree 1, i.e. all closed points P defined over $\mathbb{F}_q$ and not in $\text{supp}\,\mathscr{F}$. Such divisors are in bijective correspondence with the polynomials $P_D(t) = t - a$, $a \in \mathbb{F}_q$, and the associated function is

$$f_D\left(\frac{1}{T}\right) = P_D\left(\frac{1}{T}\right) = \frac{1}{T} - a;$$

therefore

$$\omega\left(f_D\left(\frac{1}{T}\right)\right) = \lambda(a);$$

we also have

$$f_D(\mathscr{F}_0) = \prod_v (\xi_v - a)^{a_v} = (-1)^m f_{\mathscr{F}_0}(a), \qquad m = \sum_v a_v;$$

hence

$$\sum_{a}{}' \lambda(a)\chi(f_{\mathscr{F}_0}(a)) = \chi(-1)^m(-1)\sum_{i=1}^{r} \alpha_i.$$

where $\sum_a'$ is taken over all closed points P corresponding to polynomials $T - a$, $a \in \mathbb{F}_q$, except those, if any, contained in supp $\mathscr{F}_0$. We can extend the sum to all closed points defined over $\mathbb{F}_q$ by agreeing that $\chi(0) = \chi(\infty) = 0$. We have thus proved the following result.

Theorem 4.5 *With the assumptions and notation as above we have*

$$\left| \sum_a \lambda(a)\chi(f_{\mathscr{F}_0}(a)) \right| \le (N - d - 2)q^{1/2}.$$

Remarks Beyond the basis estimates provided by the Riemann hypothesis for function fields, there are two other results used in the proof of the above theorem:

1. The function Λ is an Abelian character, i.e. there exists an Abelian cover $\mathscr{C}/\mathbb{P}^1$ with Galois group $\text{Gal}(\mathscr{C}/\mathbb{P}^1)$ admitting Λ as a character,

2. The zeta function $Z(s, \mathscr{C})$ of $\mathscr{C}$ admits a decomposition as a product of L-functions $L(s, \Lambda, \mathbb{P}^1)$ with Λ ranging over all the characters of $\text{Gal}(\mathscr{C}/\mathbb{P}^1)$.

These two statements are at the heart of the development of class field theory for function fields in one variable with finite field of constants. A proof along classical lines is given by F. K. Schmidt in Die Theorie der Klassenkorper uber einem endlichen Koeffzienten bereich, *Sitz.-Ber. Erlangen*, 62(1930), 267–284. The proof is modeled on Takagi's proof for number fields and uses the analytic techniques already developed in Chapter 3. As there are no Archimedean primes in the function field case, the core of the proof reduces to some straightforward index calculations. The particular situation dealt with in this section refers to the class field theory of the rational function field $\mathbb{F}_q(T)$ of the projective line $\mathbb{P}^1$. As an analogue of Kronecker's theorem which implies that cyclotomic fields are the class fields of the rationals $\mathbb{Q}$, it is possible to develop an explicit class field theory for $\mathbb{F}_q(T)$, where the role of the cyclotomic equations is now played by some interesting polynomials studied by Carlitz. These ideas are very clearly presented in the paper by D. R. Hayes, Explicit class field theory for the rational function field, *Trans. A. M. S.*, **189** (1974), 77–91. When one of the characters χ or ω is trivial, the corresponding covering $\mathscr{C} \to \mathbb{P}^1$ is cyclic and the basic theory can be treated in a direct way using either Kummer theory or Artin–Schreier theory applied to $\mathbb{F}_q(T)$. This is done by H. Hasse in, Theorie der relative-zyklischen algebraischen Funktionen-korper, insbesondere bei endlichen Konstantenkorper (*Crelle*, **172** (1934), 37–54). Class field theory for an arbitrary function field in one variable with finite field of constants is dealt with in E. Artin & J. Tate, *Class Field Theory* (W. A. Benjamin, New York, 1967). The modern geometric theory has been developed by Lang and Serre and is very lucidly presented in J.-P. Serre's book *Groupes Algébriques et les Corps de Classes* (Herman, Paris, 1959). Some interesting applications to exponential sums of the theory of *l*-adic representations have been made by Bombieri, Deligne and Katz. This aspect of the theory is developed in N. Katz's monograph, Sommes Exponentielles (*Astérisque*, **79** (1980)).

Example If $\psi: \mathbb{F}_q^+ \to \mathbb{C}^\times$ is a non-trivial additive character, we can define a character ω of conductor (T^{-2}) by putting for every series in U_1

$$\omega(1 + a_1 T + a_2 T^2 + \cdots) = -\psi(a_1).$$

The Weil estimate gives

$$\left| \sum_a \psi(a)\chi(f_{\mathscr{F}_0}(a)) \right| \leq dq^{1/2}.$$

If for the function $f_{\mathscr{F}_0}$ we take a quadratic polynomial

$$f_{\mathscr{F}_0} = t^2 + \partial t + \gamma$$

with non-zero discriminant and char $\mathbb{F}_q \neq 2$, then

$$\left| \sum_a \psi(a)\chi(a^2 + \beta a + \gamma) \right| \leq 2q^{1/2}.$$

The case where χ is the unique character $\mathbb{F}_q^\times$ of order 2 leads after an elementary transformation, say in the case $\mathbb{F}_q = \mathbb{F}_p$, to the Kloosterman sum already introduced in Section 4.4:

$$\sum_{a=1}^{p-1} e^{2\pi i(ca+bc^{-1})/p},$$

with suitable b and c.

A procedure for constructing more general exponential sums, of the type which are useful in number theory, is provided by the following elementary lemma.

Lemma 4.9 (Weil) *Let $\psi: \mathbb{F}_q^+ \to \mathbb{C}^\times$ be a non-trivial additive character. Let f be a polynomial in $\mathbb{F}_q[T]$ of degree d with $f(0) = 0$. Then there is a character $\omega: U_1 \to \mathbb{C}^\times$ of order p, of conductor (T^{-N}) with some $N \leq d + 1$, such that*

$$\omega(1 + aT^{-1}) = \psi(f(a))$$

for all $a \in \mathbb{F}_q$.

Proof. It suffices to prove the result for the polynomial $f = aT^d$ with $d \neq 0$, $a \in \mathbb{F}_q^\times$. Let $X_1, X_2, \ldots$, and U be indeterminates; in the ring of formal power series $Q[X_1, X_2, \ldots][U]$ consider the identity

$$\frac{d}{dU} \log(1 + V) = (1 + V)^{-1} \frac{dV}{dU} = \sum_{n=1}^{\infty} P_n U^{n-1},$$

where

$$V = \sum_{n=1}^{\infty} X_n U^n.$$

It is easy to verify that

$$P_n \in Z[X_1, X_2, \ldots, X_n] \qquad \text{for all } n \geq 1 \tag{4.5}$$

$$P_n(X_1, 0, \ldots, 0) = (-1)^{n-1} X_1^n. \tag{4.6}$$

Consider new indeterminates $Y_1, Y_2, \ldots$, and the element

$$W = \sum_{n=1}^{\infty} Y_n U^n;$$

if we put

$$(1 + V)(1 + W) = 1 + \sum_{n=1}^{\infty} Z_n U^n.$$

then we have

$$Z_n \in Z[X_1, Y_1, X_2, Y_2, \ldots]$$

$$P_n(Z_1, \ldots, Z_n) = P_n(X_1, \ldots, X_n) + P_n(Y_1, \ldots, Y_n). \tag{4.7}$$

Now if K is any field and

$$K(U)_1^\times = \{1 + c_1 U + c_2 U^2 + \cdots : c_i \in K\},$$

we obtain from the above identities a homomorphism

$$\Omega: K(U)_1^\times \to K^+$$

defined by

$$\Omega(1 + c_1 U + c_2 U^2 + \cdots) = (-1)^{n-1} a P_n(c_1, \ldots, c_n),$$

$a \in K^\times$ so that

$$\Omega(1 + c_1 U) = a c_1^n.$$

The lemma now follows by taking $K = \mathbb{F}_q$, $U = T^{-1}$ and $\omega = \psi \circ \Omega$.

Remarks 1. There are conditions that can imposed on the degree d of the polynomial f so that the conductor of ω is exactly $d + 1$. For example a simple calculation involving Gauss sums shows that when $p \equiv 1 \bmod k$, the character ω with $\omega(1 + aT^{-1}) = e^{2\pi i a^k/p}$ has conductor $k + 1$; in fact if χ is a character of order k of $\mathbb{F}_q^\times$ and $G(k) = \sum_{n=0}^{p-1} e^{2\pi i n^k/p}$, $G(\chi) = \sum_n \chi(n) e^{2\pi i n/p}$, then

$$G(k) = \sum_n e^{2\pi i n^k/p} \{1 + \chi(n) + \cdots + \chi(n)^{k-1}\} = \sum_{j=1}^{k-1} G(\chi^j):$$

from the properties of the Gauss sums $G(\chi^j)$ we know that $G(k)$ is the sum of $k - 1$ complex numbers each with absolute value $p^{1/2}$. Hence from Weil's estimate we get that

$$k - 1 \le \text{conductor}(\omega) - 2,$$

that is conductor $(\omega) = k + 1$. Actually the condition that d be prime to p suffices.

2. Besides the explicit examples provided earlier we have in the odd characteristic case a character of exponential conductor 5 given by $\omega(1 + a_1 T^{-1} + \cdots + a_4 T^{-4} + \cdots) = \psi(P_4(a_1, a_2, a_3, a_4))$, where

$$P_4(x_1, x_2, x_3, x_4) = 2x_1^2 x_2 - 4x_1 x_3 - x_1^4 + 4x_4 + 2x_2(x_1^2 - x_2).$$

Clearly $P_4(a_1, 0, 0, 0) = -a_1^4$.

4.6.2 Part II: Artin–Schreier coverings

The aim of this section is to develop the basic theory of Artin–Schreier coverings and the associated theory of exponential sums which generalize those sums of the form $\sum_x \exp(2\pi i f(x)/p)$ with $f(x)$ a rational function. The main result obtained gives an upper bound for such sums which is best possible. The ideas involved are originally due to Davenport. Hasse and Weil. The presentation given here is based on the idea that the calculation of the conductor of an Artin–Schreier covering can be reduced via localization to a similar calculation for an Artin–Schreier covering of the projective line. This is a beautiful idea of Bombieri which amply justifies the well-known principle that in a neighborhood of a point, a complete smooth algebraic curve looks like the projective line.

Let $k = k(\mathscr{C})$ be the function field of a curve with exact field of constants $\mathbb{F}_q$, $q = p^f$; Let K/k be a normal extension of degree p. Since K is separable it can be generated by the adjunction of a single element, and a generator can be chosen which satisfies a simple equation. Recall that the separability of K/k implies that $\text{Tr}_{K/k} \not\equiv 0$; let $\theta \in K$ with $\text{Tr}_{K/k}(\theta) = b \ne 0$. As θ is not an element in k, we have $K = k(\theta)$. Let $\text{Gal}(K/k) = \{\sigma^d : d = 0, \ldots, p - 1\}$. The two numbers

$$\beta = \theta + 2\sigma\theta + 3\sigma^2\theta + \cdots + (p-1)\sigma^{p-2}\theta,$$

$$\sigma\beta = \sigma\theta + 2\sigma^2\theta + \cdots + (p-2)\sigma^{p-2}\theta + (p-1)\sigma^{p-1}\theta$$

satisfy

$$\beta - \sigma\beta = \text{Tr}_{K/k}(\theta) = b \neq 0.$$

Let $\alpha = -\beta/b$; then $\sigma\alpha \neq \alpha$ and α does not lie in k and therefore, as claimed, $K = \mathbb{F}_q(\alpha)$. Also observe that $\sigma\alpha = \alpha + 1$. Hence the action of $\text{Gal}(K/k)$ is given simply by

$$\sigma^i(\alpha) = \alpha + i, \qquad i \in F_p^+ = \{0, 1, \ldots, p-1\}.$$

As for the irreducible equation satisfied by α we observe that

$$\sigma(\alpha^p - \alpha) = (\sigma\alpha)^p - \sigma\alpha = (\sigma\alpha + 1)^p - (\alpha + 1) = \alpha^p - \alpha.$$

Hence $\alpha^p - \alpha = a$ where $a \in k$. Therefore α is a root of the equation

$$x^p - x = a;$$

the other roots of this equation are $\alpha + i, i = 1, \ldots, p-1$. Conversely any polynomial in $k[x]$ of the form $f = x^p - x - a$ is either irreducible, or the product of p linear factors; for if c is a root then so are the numbers $c + i$, $i = 0, \ldots, p-1$, and hence a has the form $a = c^p - c$. If f is irreducible then the splitting field $K = K_f$ of f is a cyclic extension of degree p and $\text{Gal}(K/k) = \{\sigma^i : 0 \leq i \leq p-1\}$ with $\sigma(\alpha) = \alpha + 1$.

Suppose that k is a local field of characteristic p with discrete valuation v, and order function ord_v. Let T be a local uniformizing parameter. We want to simplify further the choice of generator in this case. Suppose then that $K = k(\alpha) = k(\beta)$ with generators α, β which satisfy

$$\alpha^p - \alpha = a, \quad \beta^p - \beta = b.$$

Let $\sigma \in \text{Gal}(K/k)$ be a generator with $\sigma\alpha = \alpha + 1$ and let $\sigma\beta = \beta + r$ with $0 < r - 1$, so that $\sigma(\beta/r) = \beta/r + 1$. Let $c = \beta/r - \alpha$ and observe that $= \sigma(\beta/r) - \sigma\alpha = (\beta/r + 1) - (\alpha + 1) = c$; hence $c \in k$. On the other hand

$$c^p - c = (\beta/r - \alpha)^p - (\beta/r - \alpha) = \beta^p/r - \alpha^p - \beta/r + \alpha = b/r - a.$$

Put $b = r(a + c^p - c)$ and observe that $\beta = r(\alpha + c)$. Assuming that k is complete under the valuation ord_v, we can write $K = k(\alpha)$ with α satisfying $\alpha^p - \alpha = a$ and a an element in k with formal power series expansion

$$a = \sum_{\mu=-m}^{\infty} c_\mu T^\mu, c_\mu \in \mathbb{F}_{q'}.$$

We now show how to use the above transformations to replace a by an element with a simpler expansion. First we modify the element a by subtracting those terms in the T-expansion corresponding to negative multiples of p, say $\mu = p\nu$: since $\mathbb{F}_q$ is perfect, if the term $c_{p\nu}T^{p\nu}$ appears in the T-expansion of a, we can subtract it without affecting the structure of $K = k(\alpha)$; in fact the equation $\alpha^p - \alpha = a$ with $a \mapsto a - (c_{p\nu}^{1/p}T^\nu)^p + (c_{p\nu}^{1/p}T^\nu)$ also generates K. Iterating this procedure a finite number of times replaces a by another element whose T-expansion contains no terms corresponding to negative multiples of p:

$$a' = b + \omega + \sum_{\nu=1}^{m} d_\nu T^{-\nu},$$

where $d_\nu = 0$ if $p|\nu$ and $\omega \in \mathbb{F}_{q'}$, $\mathrm{ord}_\nu(b) > 0$. In particular p does not divide m. Let

$$c = b + b^p + b^{p^2} + \cdots;$$

since $\mathrm{ord}_\nu(b) > 0$, this series converges to an element c which satisfies $c^p - c = -b$. Therefore a' may be replaced by

$$a'' = a' + c^p - c = \omega + \sum_{\nu=1}^{m} d_\nu T^{-\nu}.$$

Any further changes can only be effected on the constant ω. Thus the element a can be replaced by one whose T-expansion contains no positive powers of T. In particular if $\mathrm{ord}_\nu(a) \geq 0$, then the extension $K = k(\alpha)$ with $\alpha^p - \alpha = a$ is unramified, i.e. $K = k(\alpha)$ and $\alpha^p - \alpha = a$ with a constant. We have thus proved the following lemma.

Lemma 4.10 *Let $K = \mathbb{F}_q(\mathscr{C})$ be the function field of a smooth complete curve $\mathscr{C}$ and let $f \in K$. The Artin–Schreier extension $K = K(\alpha)$ with $\alpha^p - \alpha = f$ is unramified outside the poles of f.*

Let $k = \mathbb{F}_q$ be the finite field of $q = p^f$ elements. Let $\mathscr{C}$ be a complete non-singular curve of genus g defined over k with function field $K = k(\mathscr{C})$ and let $\bar{k}(\mathscr{C})$ be the function field of $\mathscr{C}$ considered as a curve over the algebraic closure $\bar{k}$ of $\mathbb{F}_q$.

Let $f \in k(\mathscr{C})$ be a rational function on $\mathscr{C}$ and suppose the equation $z^p - z - f(x) = 0$ is absolutely irreducible over $\bar{k}(\mathscr{C})[z]$, i.e. $z^p - z \neq f(x)$ for any $z \in \bar{k}(\mathscr{C})$. At a point P_ν of degree d with local uniformizing parameter T_ν and exponential valuation ord_ν, we say that f has a pole if $\mathrm{ord}_\nu(f) < 0$; otherwise the value of f at P_ν is defined to be a_0, where

$$f = \sum_{i \geq 0} a_i T_\nu^i, \qquad a_i \in \mathbb{F}_{q^d}.$$

For notational convenience we shall often write the value of f at P_v as $f(P_v)$.

Let $\tilde{K}$ be the normal extension of $K = k(\mathscr{C})$ obtained by adjoining the roots of the equation $z^p - z = f(x)$. This is a Galois extension of the type considered above with Galois group $\text{Gal}(\tilde{K}/K) = \{0, 1, \ldots, p - 1\}$, where the action is given by $\sigma(z) = z + 1$. Let $\tilde{\mathscr{C}}$ be a smooth model for the function field K; the corresponding covering $\pi \colon \tilde{\mathscr{C}} \to \mathscr{C}$ is called the Artin–Schreier covering associated with the function f. It is clear that both $\tilde{\mathscr{C}}$ and $\mathscr{C}$ are defined over F_q.

If k_v is a finite extension of $k = \mathbb{F}_q$ with absolute trace $\text{Tr} = \text{Trace}_{k_v/\mathbb{F}_p}$, then we define as before a character

$$\psi_{k_v}(x) = \exp\{2\pi i \, \text{Tr}(x)/p\}.$$

We now define a multiplicative character on the group of divisors of $\mathscr{C}$ whose support is disjoint from the poles of f by putting for a closed point P_v

$$\Lambda(P_v) = \psi_{k_v}(f(P_v)),$$

where k_v is the residue class field of P_v. Let $(f)_\infty$ denote the divisor of poles of $f(x)$ on $\mathscr{C}$ and write

$$(f)_\infty = -\sum_v \text{ord}_v(f)P_v.$$

where the sum $\sum_v$ is taken only over the poles P_v of f on $\mathscr{C}$.

Definition 4.5 Let $d_v = \text{ord}_v(f)$ and define a divisor

$$\mathscr{F} = \sum_v f_v P_v \tag{4.8}$$

on $\mathscr{C}$ by prescribing that the multiplicities f_v be given by $f_v = d_v + 1$ if P_v is a pole of f and $f_v = 0$ otherwise.

As before, we associate to the character Λ the L-function

$$L(s, \Lambda, \mathscr{C}) = \prod_P' \frac{1}{1 - \Lambda(P)NP^{-s}},$$

where the product $\prod_P'$ is taken over all closed points P on $\mathscr{C}$ which are not poles of f. The main result of this section is the following theorem.

Theorem 4.47 *With notation and assumptions as above we have:*

(i) *At a pole P_v of f of multiplicity d_v, the local conductor of the character Λ is $\leq f_v = d_v + 1$.*

(ii) *As a function of $u = q^{-s}$, the L-function $L(s, \Lambda, \mathscr{C})$ is a polynomial of degree*

$$\deg L(s, \Lambda, \mathscr{C}) \le 2g - 2 + \deg \mathscr{F}.$$

with equality holding if and only if $(d_v, p) = 1$ for all poles P_v of f.

(iii)   *The exponential sum*

$$S(f) = \sum_{P}' \psi_k(f(P)),$$

where the sum $\sum_P'$ is taken over all points P on $\mathscr{C}$ of degree one which are not poles of f satisfies

$$|S(f)| \le (2g - 2 + \deg \mathscr{F})q^{1/2}.$$

Proof. (ii) is a consequence of the definition of the divisor $\mathscr{F}$ and the results proved in Chapter 3 concerning the nature of the polynomial $L(s, \Lambda, \mathscr{C})$. As for (iii) we recall that, with $u = q^s$, $L(s, \Lambda, \mathscr{C})$ is a polynomial whose linear term is given by

$$L(s, \Lambda, \mathscr{C}) = 1 + S(f)u + \cdots.$$

Hence by the Riemann hypothesis applied to the zeta function of $\widetilde{\mathscr{C}}$, of which $L(s, \Lambda, \mathscr{C})$ is a factor (see below), we get that $S(f)$ is a sum of at most $2g - 2 + \deg \mathscr{F}$ complex numbers α_i with absolute value $|\alpha_i| = q^{1/2}$.

Let $Q(\zeta_p)$ be the cyclotomic extension obtained by adjoining $\zeta_p = e^{2\pi i/p}$ to Q and observe that $L(s, \Lambda, \mathscr{C})$ is a polynomial in $u = q^{-s}$ with coefficients in $Q(\zeta_p)$. If $E = \deg L(s, \Lambda, \mathscr{C})$, then the degree of any other conjugates of $L(s, \Lambda, \mathscr{C})$ is also E. If $\mathrm{Gal}(Q(\zeta_p)/Q) = (\mathbb{Z}/p\mathbb{Z})^{\times} \equiv \{1, 2, \ldots, p-1\}$, then under the correspondence $\sigma \mapsto i$, $1 \le i \le p - 1$, with $\sigma(\zeta_p) = \zeta_p^i$, we have

$$\sigma L(s, \Lambda, \mathscr{C}) = L(s, \Lambda^i, \mathscr{C}).$$

From the identity

$$\prod_{i=0}^{p-1} (1 - \Lambda^i(P)t) = \begin{cases} (1 - t)^p, & \text{if } \Lambda(P) = 1 \\ (1 - t^p), & \text{if } \Lambda(P) \ne 1 \end{cases}$$

we obtain that

$$\mathrm{Nr}_{Q(\zeta_p)/Q} L(s, \Lambda, \mathscr{C}) = Z(s, \widetilde{\mathscr{C}})/Z(s, \mathscr{C}), \tag{4.9}$$

where the norm is from $Q(\zeta_p)$ to Q, the automorphisms acting trivially on the variable $u = q^{-s}$, and $Z(s, \widetilde{\mathscr{C}})$, (resp. $Z(s, \mathscr{C})$) is the zeta function of $\widetilde{\mathscr{C}}$ (resp. $\mathscr{C}$). From Chapter 3 we know that if $\mathscr{C}$ is a complete non-singular curve of genus g then $L(s, \Lambda, \mathscr{C})$ is a rational function of $u = q^{-s}$ of degree $2g - 2$; thus (4.9) gives

$$(p - 1)E = (2\tilde{g} - 2) - (2g - 2),$$

where $\tilde{g}$ denotes the genus of $\widetilde{\mathscr{C}}$ and g that of $\mathscr{C}$.

We shall now use Hilbert's version of the Hurwitz-Zeuthen formula to estimate the genus $\tilde{g}$ of $\tilde{\mathscr{C}}$ in terms of the higher ramification groups of the Artin–Schreier covering $\pi\colon \tilde{\mathscr{C}} \to \mathscr{C}$. This will provide an estimate for the degree of the global conductor $\mathscr{F}$ of Λ. The main result, as stated in (i) of the theorem, is due to Hasse. The proof given here, due to Bombieri ([7]), reduces the problem to a local question on the projective line, where one can argue directly using the theory of the elementary symmetric functions.

4.6.3 The Hurwitz–Zeuthen formula for the covering $\pi\colon \tilde{\mathscr{C}} \to \mathscr{C}$

As the function field of $\tilde{\mathscr{C}}$ and $\mathscr{C}$ both have $\mathbb{F}_q$ as the exact field of constants, the formula in question is a simple consequence of the relation between the canonical class in $\tilde{\mathscr{C}}$ and the canonical class in $\mathscr{C}$, i.e. the relation between the degree of the divisor of a differential on $\mathscr{C}$ as compared to its degree as a differential on $\tilde{\mathscr{C}}$.

Let $P_w \in \tilde{\mathscr{C}}$ be a closed point corresponding to the discrete valuation ord_w. Let $\hat{K}_w$ be the completion of $\tilde{K}$ with respect to ord_w and let k_w be the residue class field. For a point $P_v \in \mathscr{C}$ we also denote the corresponding valuation by ord_v and let K_v be the completion of K with respect to ord_v and let k_v denote the residue class field.

Since the Galois group $G = \mathrm{Gal}(\tilde{\mathscr{C}}/\mathscr{C})$ of the covering $\pi\colon \tilde{\mathscr{C}} \to \mathscr{C}$ is cyclic of order p, the decomposition group

$$D(w) = \{\sigma \in \mathrm{Gal}(\tilde{\mathscr{C}}/\mathscr{C})\colon \sigma(P_w) = P_w\}$$

of a point $P_w \in \tilde{\mathscr{C}}$ is either the identity 0 or all of G. Let $P \in \mathscr{C}$ and suppose e is the ramification index of P_v, g is the number of distinct points $P_w \in \tilde{\mathscr{C}}$ which lie above P_v, f is the degree of a point P_w above P_v, then we have $p = egf$; hence only the following possibilities may arise for the triple $(e, f, g)\colon (p, 1, 1), (1, p, 1), (1, 1, p)$. This implies that the only case with $D(w) = \mathrm{Gal}(\tilde{\mathscr{C}}/\mathscr{C})$ occurs when P_w lies above a point $P_v \in \mathscr{C}$ which ramifies completely in the covering $\pi\colon \tilde{\mathscr{C}} \to \mathscr{C}$. As we saw earlier this covering can ramify only at the poles P_v of $f(x)$ and if $(p, d_v) = 1$, then ramification does occur with ramification index $e = p$. Recall that if the point $P_w \in \tilde{\mathscr{C}}$ lies above $P_v \in \mathscr{C}$ then the extension $\hat{K}_w/K_v$ is Galois with $\mathrm{Gal}(\tilde{K}_w/K_v) = D(w)$, the decomposition group. The i-th ramification group of P_w is the subgroup of $D(w)$ defined by

$$D_i(w) = \{\sigma \in D(w)\colon \mathrm{ord}_w(T_w^\sigma - T_w) \geq i\}.$$

The formula in question is given by

$$(2\tilde{g} - 2) = p(2g - 2) + \sum_w \sum_{i \geq 0} (\mathrm{card}\, D_i(w) - 1), \qquad (4.10)$$

where the sum $\sum_w$ is taken only over the points P_w lying above the poles of $f(x)$ and $\mathrm{card}\, D_i(w)$ is the order of $D_i(w)$.

Suppose the covering $\pi\colon \tilde{\mathscr{C}} \to \mathscr{C}$ is ramified at the pole $P_v \in \mathscr{C}$ of $f(x)$ with $\mathrm{ord}_v(f) = -d$; let $P_w \in \tilde{\mathscr{C}}$ be the point above P_v. Since $\mathrm{Gal}(\tilde{\mathscr{C}}/\mathscr{C})$ is cyclic of order p, the ramification group $D_i(w)$ can only be $\mathrm{Gal}(\tilde{\mathscr{C}}/\mathscr{C})$ or 0; now by definition $D_i(w) = \mathrm{Gal}(\tilde{\mathscr{C}}/\mathscr{C})$ if and only if

$$i \le \mathrm{ord}_w(T_w^\sigma - T_w), \ \sigma \ne 0,$$

where T_w is a local uniformizing parameter at P_w, σ is a generator of $\mathrm{Gal}(\tilde{\mathscr{C}}/\mathscr{C})$ and where T_w^σ denotes the action of σ on T_w. This implies that the number of i with $D_i(w) = \mathrm{Gal}(\mathrm{Cal}\,\mathscr{C}/\mathscr{C})$ is $\mathrm{ord}_w(T_w^\sigma - T_w) + 1$ and for each of these we have card $D_i(w) = p$. Hence from (4.10) we obtain

$$(2\tilde{g} - 2) \le p(2g - 2) + \sum_w (\mathrm{ord}_w(T_w^\sigma - T_w) + 1)(p - 1),$$

where $\sum_w$ is taken over the points P_w which lie above the poles of $f(x)$, i.e. $P_w \in \mathrm{supp}\,\mathscr{F}$. Observe that equality holds above if and only if $\pi\colon \tilde{\mathscr{C}} \to \mathscr{C}$ is ramified at all poles of $f(x)$. As remarked earlier this is indeed the case if p does not divide $\mathrm{ord}_w(f)$ for all P_w above the poles of $f(x)$.

The proof of the theorem has now been reduced to verifying that for a generator $\sigma \in \mathrm{Gal}(\tilde{\mathscr{C}}/\mathscr{C})$ and a point P_w as above with $\mathrm{ord}_w(f) = -d$ we have

$$\mathrm{ord}_w(T_w^\sigma - T_w) \le d, \qquad (4.11)$$

with equality holding if and only if $(d, p) = 1$.

The above problem is of a local nature and concerns the ramification properties of the local field extension $\tilde{K}_w/K_v$ generated by the Artin–Schreier equation $z^p - z = a$ with a an element in K_v which satisfies $\mathrm{ord}_v(a) = -d$. We are now going to exploit the useful fact that in the neighborhood of a simple point the curve $\mathscr{C}$ looks like the projective line $\mathbb{P}^1$. As we can think of the point P_v as 'a point at infinity', we can introduce a local uniformizing parameter T^{-1} for the valuation ord_v so that the local field K_v is isomorphic to the field $F_v((T^{-1}))$ of formal Laurent series in T^{-1} with coefficients in the residue class field F_v of P_v and with a finite number of negative powers of T^{-1}. Furthermore we choose T so that if $(d, p) = 1$, then $a = T^d$, otherwise, when $(d, p) = p$ we let $a = P(T)$, where $P(T)$ is a polynomial in T of degree d. In the present setting we can think of $\mathscr{C}$ as the projective line $\mathbb{P}^1$ over F_v and the function f is a polynomial in T of degree d. The covering $\pi\colon \tilde{\mathscr{C}}' \to \mathbb{P}^1$ is defined as before by the equation

$$z^p - z = T^d, \qquad (d, p) = 1$$

or by

$$z^p - z = P(T), \qquad (d, p) = p.$$

Here the genus g of $\mathbb{P}^1$ is 0 and the only pole of f is $P_v = \infty$. Applying (4.9) and (4.10) to $\tilde{\mathscr{C}}' \to \mathbb{P}^1$ we obtain

$$(p - 1)E' = 2\tilde{g}', \qquad (4.9')$$

$$2\tilde{g}' - 2 = -2p + (p - 1) + \text{ord}_{w'}(T_{w'}^\sigma - T_{w'})(p - 1), \qquad (4.10')$$

where $\tilde{g}'$ is the genus of $\tilde{\mathscr{C}}'$, $E' = \deg L(s, \Lambda, \mathbb{P}^1)$ with Λ the character of $\text{Div}(\mathbb{P}^1 - \{\infty\})$ associated to $f = T^d$ if $(p, d) = 1$ or $f = P(T)$ if $(p, d) = p$; $P_{w'}$ is the point of $\tilde{\mathscr{C}}'$ lying above the pole $P_v = \infty$ of T^d or $P(T)$.

As the local behavior of the covering $\pi: \tilde{\mathscr{C}} \to \mathscr{C}$ at P_w is the same as that of $\tilde{\mathscr{C}}' \to \mathscr{C}$ we have

$$\text{ord}_w(T_w^\sigma - T_w) = \text{ord}_{w'}(T_{w'}^\sigma - T_{w'});$$

to complete the proof of (4.11) and hence of the theorem we observe that the statement

$$\text{ord}_{w'}(T_{w'}^\sigma - T_{w'}) \le d,$$

with equality if and only if $(d, p) = 1$, is equivalent to the statement: $E' \le d - 1$ with equality if and only if $(d, p) = 1$, i.e. the character $\Lambda(P_v) = \psi_{F_v}(f(P_v))$, with $f = T^d$ if $(d, p) = 1$ and $f = P(T)$ if $(d, p) = p$, has *conductor* $\le d + 1$ with equality if and only if $(d, p) = 1$. This is the content of the following lemma.

Lemma 4.11 *Let* $\pi: \tilde{\mathscr{C}} \to \mathbb{P}^1$ *be the Artin–Schreier covering defined by the equation* $Y^p - Y = f(X)$, *with* $f(x)$ *a polynomial in* $\mathbb{F}_q[X]$ *of degree* d. *Let* $\Lambda: \text{Div}(\mathbb{P}^1 - \{\infty\}) \to \mathbb{C}^\times$ *be the Abelian character defined for a closed point* P *by* $\Lambda(P) = \psi_{k(P)}(f(P))$ *and by multiplicativity for other divisors. Then as a polynomial in* $t = q^{-s}$ *we have*

$$\deg L(s, \Lambda, \mathbb{P}^1) \le d - 1,$$

i.e. conductor$(\Lambda) \le d + 1$, *with equality holding if and only if* $(d, p) = 1$.

Proof. We have already seen in Section 4.6 that conductor$(\Lambda) \le d + 1$ provided $Y^p - Y = f(X)$ is absolutely irreducible. Furthermore if h is any polynomial defined over F_q then the character defined by $\Lambda'(P) = \psi_{k(P)}(f(P) + h(P)^p - h(P))$ defines the same character as Λ. Therefore if $(d, p) = p$ then conductor$(\Lambda) < d - 1$. It remains to deal with the case of equality.

From the Euler identity

$$L(s, \Lambda, P^1) = \prod_P \frac{1}{1 - \Lambda(P)NP^{-s}}$$

$$= \sum_{m=0}^{E} A_m t^m$$

we have

$$A_m + \sum_{D}' \Lambda(D),$$

where the sum $\sum_D'$ is taken over all finite positive divisors D defined over F_q with $\deg(D) = m$. As such divisors

$$D = \sum_P m_P P$$

are in one-to-one correspondence with the polynomials

$$G(x) = \prod_{i=1}^{m} (x - \lambda_i),$$

we have

$$A_m = \sum_{G}' \psi_{F_0} \left(\sum_{i=1}^{m} f(\lambda_i) \right), \qquad (4.12)$$

where $\sum_G'$ is taken over all monic polynomials $G(x)$ in $\mathbb{F}_q[x]$ of degree m

$$G(x) = x^m + y_1 x^{m-1} + \cdots + y_m.$$

Observe that the argument of ψ_{F_q} is an element of $\mathbb{F}_q$, we now take $m = d - 1$; using Newton's and Waring's formulas for the elementary symmetric functions we can express $\sum_i f(\lambda_i)$, $f(x) = a_0 x^d + \cdots + a_d$, as a symmetric function of the roots of $G(x) = 0$:

$$\sum_{i=1}^{m} f(\lambda_i) = a_0 \sum_{1 \le m < d/2}'' y_{d-m}(m y_m + P_m(y_1, \ldots, y_{m-1})) + Q(y_1, \ldots, y_{[d/2]}),$$

where P_m is a polynomial in $y_1, \ldots, y_{m-1}$, $(y_0 = 1)$, and Q is a polynomial in $y_1, \ldots, y_{[d/2]}$, which is quadratic in $y_{[d/2]}$ if d is even. It is clear that the sum $\sum_m''$ is a linear form in the coefficients $y_d, y_{d-1}, \ldots, y_{d'}$, where d' is the smallest positive integer $\le d/2$. Substituting (4.10) in (4.12) we see that the sum over y_{d-1} is 0 unless the coefficient of y_{d-1} vanishes. As this coefficient is $a_0 y_1 + b_1$ with some constant b_1, we see that the sum over y_{d-1} is 0 if $a_0 y_1 + b_1 \ne 0$. Since $a_0 \ne 0$, we let $y_1 = -b_1/a_0$ and observe that the contribution $\sum y^{d-1}$ in (4.12) is q. The term $\sum_m''$ with $m = 2$ is again 0 unless $a_0(2y_2 + P_2(-b_1/a_0)) = 0$ and so on. If d is odd we continue in this way up to $m = (d-1)/2$ in which case the total contribution to (4.12) is $q^{(d-1)/2}$ times a p-th root of unity. If d is even then we continue up to $m = d/2 - 1$, thus getting a contribution of $q^{(d/2)-1}$ times a Gauss sum resulting from the quadratic term $y_{d/2}^2$ which appears in Q. As the absolute value of the Gauss sum is $q^{1/2}$, we get in both cases that the total contribution to (4.12) when the y_m's range over all values in $\mathbb{F}_q$ is

$$A_{d-1} = W(f)q^{(d-1)/2},$$

where $W(f)$ is a p-th root of unity. This proves the lemma and hence the proof of the theorem is complete.

Remarks An inductive proof of (4.12) can be based on the following three formulas: Let $g(T) = T^D + y_1 T^{D-1} + \cdots + y_D = (T - \alpha_1) \ldots (T - \alpha_D)$. Let $S_m = \sum_{i=1}^D \alpha_i^m$, $(S_0 = D)$. For $m \leq D$ we have

$$S_m + y_1 S_{m-1} + \cdots + y_{m-1} S_1 + m y_m = 0. \tag{4.13}$$

For $m \geq D$ we have

$$S_m + y_1 S_{m-1} + \cdots + y_D S_{m-D} = 0 \tag{4.14}$$

$$S_m = \sum_r \frac{(-1)^r m (r_1 + \cdots + r_D - 1)!}{r_1! \ldots r_D!} y_1^{r_1} \ldots y_D^{r_D}, \tag{4.15}$$

where the sum $\sum_r$ is taken over all D-tuples $r = (r_1, \ldots, r_D)$ of non-negative integers which are solutions of the diophantine equation

$$m = 1 r_1 + 2 r_2 + \cdots + D r_D;$$

here we take $r = r_1 + \cdots + r_D$.

2. With the notation as in the proof of the lemma above we can define a map

$$\eta: \mathbb{F}_q^{d-1} \to \mathbb{F}_q, \eta(y_1, \ldots, y_{d-1}) = \sum_i f(\lambda_i).$$

For all choices of the coordinates $y_1, \ldots, y_{[d/2]-1}$, except for one, the map η is onto. In fact for $d = 2D + 1$ there is a choice of $y_1, \ldots, y_D$ such that η is constant for all values of $y_{D+1}, \ldots, y_{2D}$. For $d = 2D$ there is a choice of $y_1, \ldots, y_{D-1}$ such that η is constant for all choices of $y_{D+1}, \ldots, y_{2D-1}$ and quadratic in y_D.

Exercises

1. (Vaughan) Let $f(n)$ be an arithmetic function, i.e. defined on the integers, with support in the closed interval $[1, X]$. Let $\mu(n)$ be the Möbius function and $\Lambda(n)$ the von Mangoldt function defined to be $\log p$ when $n = p^k$ and 0 otherwise. Prove the identity:

$$\sum_n \Lambda(n) f(n) = S_1 - S_2 - S_3 + S_4,$$

where

$$S_1 = \sum_{m \le u} \sum_n \mu(m)(\log n) f(mn)$$

$$S_2 = \sum_{m \le uv} \sum_n c_m f(mn), \qquad c_m = \sum_{d \le u} \sum_{k \le v, dk=m} \mu(d)\Lambda(k)$$

$$S_3 = \sum_{m > u} \sum_{n > v} \tau_m \Lambda(n) f(mn), \qquad \tau_m = \sum_{d \mid m; d \le u}$$

$$S_4 = \sum_{n \le v} \Lambda(n) f(n).$$

2. Prove that if X, Y, α are real numbers with $X \ge 1$, $Y \ge 1$ and that $|q\alpha - a| \le q^{-2}$ with $(a, q) = 1$, then

$$\sum_{x \le X} \min(X Y x^{-1}, \|\alpha x\|^{-1}) \ll XY(q^{-1} + Y^{-1} + q(XY)^{-1})\log(2Xq),$$

where $\|\beta\| = \min_{n \in \mathbb{Z}} |\beta - n|$.

3. Suppose that $(a, q) = 1$, $q \le n$ and $|q\alpha - a| \le q^{-1}$ then

$$\sum_{p \le n} (\log p) e^{2\pi i \alpha p} \ll (\log n)^4 (nq^{-1/2} + n^{4/5} + n^{1/2}q^{1/2}).$$

(*Hint*: Use (1) and (2) above.)

4. (Carlitz) Let $q = 2^n$. Let $a \in \mathbb{F}_q^\times$ and ψ be a non-trivial additive character of $\mathbb{F}_q^+$. Define $S(a) = \sum_{x,y \in F_q^\times} \psi(x + y + ax^{-1}y^{-1})$ and $K(a) = \sum_{x \in F_q^\times} \psi(x + ax')$. Prove (1) $K(a)^2 = q + S(a)$. (2) $-q < S(a) \le 3q$.

5. (Kummer) Let p be a prime, $p \equiv 1 \bmod 3$. Define

$$z_0 = \sum_{a=0}^{p-1} e^{2\pi i x^3/p}.$$

Prove that $z_0 = \tau + \bar{\tau}$, $\tau = \sum_{n=1}^{p-1} \chi(n) e^{2\pi i n/p}$. Verify also that z_0 is a root of the cubic equation $z^3 - 3pz - pa = 0$, where $4p = a^2 + 27b^2$, $a \equiv 1 \bmod 3$.

6. (Incomplete exponential sums) Let q be an integer $m \ge 1$ and let $\varphi: \mathbb{Z}/m \to \mathbb{C}$ and extend $\varphi: \mathbb{Z} \to \mathbb{C}$ by putting $\varphi(x) = \varphi(y)$ if $x \equiv y \bmod m$. For two integers a, b put

$$S_{a,b}\varphi = \sum_{a \le \alpha < b} \varphi(x).$$

Let $\lambda \in \mathbb{Z}/m\mathbb{Z}$ and put

$$(S_\lambda \varphi)(x) = \sum_{x \bmod m} e^{2\pi i a x \lambda} \varphi(x).$$

Prove that

$$|S_{a,b}\varphi| \le (1 + \log m) \operatorname{Max}_\lambda |S_\lambda \varphi|$$

and for P a positive integer

$$\sum_{1 \le \alpha \le P} \varphi(x) = (P/m) \sum_{1 \le \alpha \le m} \varphi(x) + O(\log m \operatorname{Max}_\lambda |S_\lambda \varphi|).$$

7. Let q, a, b be integers and define the exponential sum

$$S(q,a,b) = \sum_{x=1}^{q} e^{2\pi i(ax^k+bx)/q}.$$

(i) Prove that for p a prime $S(q,a,b) \le kp^{1/2}$. (ii) Prove that as a function of q, $S(q,a,b)$ is multiplicative. (iii) A well-known theorem of Hua implies that $S(q,a,b) \ll q^{1/2+\varepsilon}(q,b)$. Use this estimate to show that for $k \ge 2$

$$\sum_{x=1}^{P} e^{2\pi iax^k/q} = (P/q)\sum e^{2\pi iax^k/q} + O(q^{1/2+\varepsilon}).$$

8. (Gauss: 'Summatio quarumdam serierum singularium.') For positive integers m and n, $0 \le m \le n$ define the rational function in the indeterminate x

$$\begin{bmatrix} n \\ m \end{bmatrix} = \prod_{i=1}^{m} (1 - x^{n+1-i})/(1 - x^i).$$

For $m = 0$ put $\begin{bmatrix} n \\ 0 \end{bmatrix} = 1$ for $n = 0, 1, 2, \ldots$, Prove

$$\begin{bmatrix} n \\ m \end{bmatrix} = \begin{bmatrix} n \\ n-m \end{bmatrix}$$

$$\begin{bmatrix} n \\ m \end{bmatrix} = \begin{bmatrix} n-1 \\ m-1 \end{bmatrix} + x^m \begin{bmatrix} n-1 \\ m \end{bmatrix}$$

$$\begin{bmatrix} n \\ m \end{bmatrix} = \begin{bmatrix} n-1 \\ m \end{bmatrix} + x^{n-m} \begin{bmatrix} n-1 \\ m-1 \end{bmatrix}$$

$$\prod_{i=1}^{n} (1 + x^{i-1}y) = \sum_{j=0}^{n} \begin{bmatrix} n \\ j \end{bmatrix} x^{j(j+1)/2} y^j.$$

The sum

$$f(x,m) = \sum_{j=0}^{m} (-1)^j \begin{bmatrix} m \\ j \end{bmatrix}$$

is 0 if m is odd and

$$f(x,2n) = \prod_{j=1}^{n} (1 - x^{2j-1}).$$

For k odd and $\zeta = e^{2\pi i/k}$ we have

$$f(\zeta^{-2}, k-1) = \zeta^{-(k+1)^2/4}G(\zeta),$$

where

$$G(\zeta) = \sum_{j \bmod k} \alpha^{j^2}.$$

Use the above identity and the fact that $|G(\zeta)|^2 = k$ to show that

$$G(\zeta) = i^{(k-1)^2/4}k^{1/2}.$$

9. (Kummer) Let a be a positive integer, let z be a primitive root of $z^{p^\nu} - 1 = 0$, p an odd prime, g a primitive root of the congruence $g^{p^{a-1}(p-1)} \equiv 1 \bmod p^a$, $\beta = \omega z^{rp}$, $\omega^{p-1} = 1$ and r an arbitrary integer. Put

$$(\beta, z) = z + \beta z^g + \beta^2 z^{g^2} + \cdots + \beta^{\phi(p^\nu)-1} z^{g^{\phi(p^\nu)-1}}.$$

Prove the following statements. (1) If $r \equiv 0 \bmod p$ then $(\beta, z) = 0$. (ii) If a is even then $(\beta, z) = p^{a/2} \omega^p z^{rp\rho + g^\rho}$, where ρ is determined by the congruence

$$r(p-1) + g^\rho \sum_{1 \le j \le a/2} (-1)^{j-1} e^j p^{j-1}/j \equiv 0 \bmod p^{a/2},$$

where $r \equiv e \bmod p$. (iii) $(\beta, z) = \pm(-1)^{(p-1)/4} p^{a/2} \omega^p z^{rp\rho + g^\rho}$. (iv) $(\beta^{-1}, z^{-1}) = p^a$.

10. Let $f(x)$ be a polynomial whose first derivative is not identically 0. Prove that

$$\left| \sum_{x=1}^{p^2} e^{2\pi i f(x)/p^2} \right| \le (\deg f - 1)p.$$

11. Let p and w be distinct primes. Let $q = p^f$, $l = w^r$ and n the smallest integer with $q^n \equiv 1 \bmod w$, i.e. n is the order of q in $\mathbb{F}_w^\times$, put $e = (w-1)/n$ and let $i_1, i_2, \ldots, i_e$ be a set of coset representatives for the group $\mathbb{F}_w^\times$ modulo the cyclic group generated by q. Let S be the set of integers defined by $w^{k-1}(cw + i_j)$ as j, k, c run over the triples satisfying the conditions $1 \le j \le e$, $1 \le k \le r$ and $1 \le c \le w^{r-k}$. Let χ be any character of $\mathbb{F}_{p^r}^\times$ and let $G_r(\chi)$ denote the corresponding Gauss sum. Prove that

$$1 = \frac{\chi(l)^l G_f(\chi)}{G_f(\chi^l)} \prod_{a \in S} \frac{G_{fn}(\tilde{\chi}\psi^a)}{G_{fn}(\psi^a)}$$

where $\tilde{\chi} = \chi^{(q^n-1)/(q-1)}$ is a character of $\mathbb{F}_{q^n}^\times$ and ψ is a character of order l of $\mathbb{F}_{q^n}$.

12. Let l and p be distinct primes and let f be the order of q in $\mathbb{F}_l$. Let T be the set of representatives for the orbits of non-trivial characters of $\mathbb{F}_{q^l}^\times$, of order l under the action of $\mathrm{Gal}(\mathbb{F}_{q^l}/\mathbb{F}_q)$. Let ρ be a character of $\mathbb{F}_q^\times$. Prove the Hasse–Davenport distribution formula:

$$G(\rho^l) \prod_{\mu \in T} G(\mu) = \rho(l^l) G(\rho) \prod_{\mu \in T} G(\mu(\rho \circ N)),$$

where N denotes the norm map $N: \mathbb{F}_{q^l} \to \mathbb{F}_q$.

13. Let $Z = KK'$ be the compositum of the extensions $K = \mathbb{F}_q(x)$, $x^m = t$ and the rational function field $K' = \mathbb{F}_q(t)$. The characters of $\mathrm{Gal}(Z/K)$ are given by the n-th power residue symbol

$$\left(\frac{(1-x^m)^\nu}{A} \right)_n, \qquad \nu = 0, 1, \ldots, n-1$$

and the associated L-functions are

$$L_\nu(s, Z/K) = \sum_A \left(\frac{(1-x^m)^\nu}{A} \right)_n \frac{1}{NA^s}, \qquad \nu = 0, 1, \ldots, n-1.$$

Use (Chapter 3, exercise 6(e) to show that

$$L_v(s, Z/K) = \prod_{\mu=0}^{m-1} L_{\mu v}(s).$$

14. Show that

$$\sum_A \left(\frac{1-x^m}{A}\right)_n \frac{1}{NA^s} = \prod_{v=0}^{m-1} (1 - \pi(\chi^\mu, \psi)q^{-s})$$

and hence that

$$\prod_{\mu=0}^{m-1} \pi(\chi^\mu, \psi) = (-1)^{m-1} \sum_A \left(\frac{1-x^m}{A}\right)_n,$$

where the sum is over all positive divisors A in $\mathbb{F}_q(x)$ of degree $m - 1$.

15. Use the reciprocity law (Chapter 3, exercise 5) to show that if the principal divisor associated with the polynomial

$$a(x) = x^{m-1} + a_1 x^{m-2} + \cdots + a_{m-1}$$

is

$$(a(x)) = A - mP_\infty$$

then

$$\left(\frac{1-x^m}{A}\right)_n = \psi\left(\prod_{\mu=0}^{m-1} *A(\zeta^\mu)\right),$$

where

$$*A(x) = 1 + a_1 x + \cdots + a_{m-1}x^{m-1} = x^{m-1}a(1/x).$$

Hence show that

$$\sum_A \left(\frac{1-x^m}{A}\right)_n = \sum_{(b_i)} \psi(b_0 b_1 \ldots b_{m-1}),$$

where $\sum_A$ is taken over all positive divisors in $\mathbb{F}_q(x)$ of degree $m - 1$ disjoint from P_∞ and $\sum_{(b_i)}$ is taken over all m-tuples $(b_0, \ldots, b_{m-1})$ in $\mathbb{F}_q^m$ satisfying $b_0 + \cdots + b_{m-1} = m$.

16. Show that

$$(-1)^{m-1}G(\psi)^m = \sum_{b_i} \psi(b_0 \ldots b_{m-1})G(\psi^m)\psi(m)^m.$$

17. Use the results of (Chapter 3, exercise 6(d)) to show that

$$\sum_a \chi^\mu(a)\psi^v(1-a) = -\pi(\chi^\mu, \psi^v).$$

where $a \in \mathbb{F}_q^x - \{1\}$.

18. (Ramanujam) Let q and n be integers with $q > 0$. Let

$$c_q(n) = \sum_{p=1, (p,q)=1}^{q} e^{-2\pi i pn/q}.$$

Show that if $(q, q') = 1$, then $c_q(n)c_{q'}(n) = c_{qq'}(n)$. Show that for $q = p^k$, $c_q(n) = 0$, if $p^{k-1} \nmid n$, $= -p^{k-1}$ if $p^{k-1} \| n$, $= p^{k-1}(p-1)$ if $p^k | n$. Show that

$$c_q(n) = \frac{\mu(q')\varphi(q)}{\varphi(q')},$$

where $\mu(q')$ is the Mobius function, $\varphi(q)$ is Euler's function and $q' = q/(q, n)$.

Notes

The modern study of exponential sums springs from three different but related sources. First there is Langlands' theory of Artin L-functions which depends fundamentally on a study of the functoriality properties of the Artin root numbers. This theory is strongly connected with the representation theory of reductive groups over local p-adic fields. In its original formulation for the group $GL(n)$, the main result concerning root numbers depends on results of Stickelberger, Hasse–Davenport, and a new relation obtained by Langlands.

The reader will find in this chapter all the necessary results concerning Gauss sums to study the Dwork–Langlands' proof of the decomposition of the global root number as a product of local root numbers (R. P. Langlands, *On the Functional Equation of the Artin L-functions*, preprint, Yale University).

The second topic concerns the study of abelian extensions of number fields in the sense of Iwasawa. A central point of the theory is the study of various relations of Stickelberger type for certain elements of ideal class groups. The theory in this direction has been pursued very actively in recent years; a noteworthy result here is the theorem of Gross and Koblitz (*Ann. of Math.*, **109** (1979)), which essentially gives the complete p-adic expansion of a power of the Gauss sum. There are also deep connections with p-adic analysis and particularly with the notion of p-adic Gamma function.

The results on abelian varieties with complex multiplications and their relation to the Taniyama group also fall in this category. The third, but not the least important, is the application of exponential sums to analytic number theory.

The most important development in this area has been the application of the theory of automorphic forms, especially the so-called Kuznietsov formula, to study the comparative behavior of Kloosterman sums for varying levels. Very deep results have been obtained by H. Iwaniec and J. Deshouillers (*Invent. Math.*, **70** (1982)) who have applied this technique to some problems in classical number theory, e.g. divisor problems. One expects that the higher dimensional analogues will yield more powerful results.

5
Goppa codes and modular curves

Simple error-correcting codes were introduced early in the development of computer hardware to aid in the detection and correction of errors. Today's revolution in the processing of information has led to the development and implementation of bigger and better codes, to the point where even the quality of the music we hear is being filtered through coding devices. Although the basic theory of algebraic codes is well understood, there remain deep theoretical problems whose solutions have escaped the efforts of the best workers of the field, e.g. the determination of the natural limits for the combined relationship between the transmission rate and the relative weight of a code, the construction of codes over the binary field with bounds which are better than the Varshamov–Gilbert bound.

Recent developments in the field of algebraic codes pioneered by Goppa seem to suggest that a unified approach to the study of some of the most important codes is best achieved from the point of view of algebraic geometry. For instance the columns of the parity matrix of a linear code over some field k can be thought of as representing points in some projective curve defined over k. For a given code there are many ways of realizing this interpretation; Goppa's idea is to exploit particular realizations. In this way one expects some interesting curves to be associated with the Golay code, to mention just one example. One aim of this chapter is to explicate Goppa's idea of constructing linear codes from the rational points on an algebraic curve. The key idea here is to use the Riemann–Roch theorem to control the transmission and relative weight rates of the codes. We have made an attempt to explain all the necessary concepts from the theory of algebraic curves with a minimum of requirements from algebraic geometry.

A major portion of the chapter is taken by a presentation of some of the fundamentals of the theory of modular curves. Our aim has been to develop the necessary tools for a complete proof of the Tsfasman–Vladut–Zink theorem concerning codes whose rates are better than the Varshamov–Gilbert bound. Our proof differs from the original in that we have followed a more direct approach to counting the rational points on modular curves via the use of Hecke operators. As a fuller treatment of the theory of modular curves would require a book several times the size of this one, we have at times only sketched the delicate arguments and recommend that the interested reader consult the many references we have given. An alternative construction of the Goppa codes has been given by van Lint [92].

5.1. Elementary Goppa codes

There are two basic definitions of Goppa codes which we want to review here.

Definition 5.1 *Let* $g(T)$ *be a (monic) polynomial of degree t over* $\mathbb{F}_{q^m}$. *Let* $L = \{\alpha_1,\ldots,\alpha_n\}$ *be a subset of n-distinct elements of* $\mathbb{F}_{q^m}$ *with* $g(\alpha_i) \neq 0$, $1 \leq i \leq n$. *The Goppa code associated with the polynomial g and the set L consists of those code words* $(c_1,\ldots,c_n)$ *in* $\mathbb{F}_q^n$ *which satisfy*

$$\sum_{i=1}^{n} \frac{c_i}{T - \alpha_i} \equiv 0 \,(\mathrm{mod}\, g(T)). \tag{5.1}$$

Definition 5.2 *Let* $g(T)$ *be a (monic) polynomial of degree t over* $\mathbb{F}_{q^m}$ *with* $g(\alpha_i) \neq 0, 1 \leq i \leq n$. *Associated with the data* $D =: \{L, g\}$, *consider the vector space* $L(D)$ *consisting of rational functions*

$$f(T)/g(T), \qquad \deg f = \deg g - 1, \qquad f \in \mathbb{F}_q[T].$$

Clearly $\dim_{\mathbb{F}_q} L(D) = t$. *Let* $\varphi_1(T),\ldots,\varphi_t(T)$ *be a fixed basis of* $L(D)$. *Define a code by means of the parity check matrix*

$$H = (\varphi_i(\alpha_j)) \qquad \text{row index} \qquad i = 1,\ldots,t$$
$$\text{column index} \quad j = 1,\ldots,n.$$

Let $w = w(c)$ denote the Hamming weight of a non-zero code word $c = (c_1,\ldots,c_n)$. Observe that the rational function representing c has, by (5.1), w poles and is therefore a quotient $A(T)/B(T)$ of polynomials in $\mathbb{F}_{q^m}[T]$ with $w = \deg B(T) = \deg A(T) - 1$. Since $A(T)$ is a non-zero multiple of g it follows that the weight of the code is $\geq t + 1$.

Now the $\mathbb{F}_q$-vector space $L(D)$ has many bases. The particular choice $\varphi_i(T) = T^i/g(T), 0 \leq i \leq t - 1$, shows readily via a Van der Monde determinant argument that the $\mathbb{F}_{q^m}$-rank of the corresponding parity check matrix

$$H = (\alpha_j^i g(\alpha_j)^{-1}), \qquad 0 \leq i \leq t - 1$$
$$1 \leq j \leq n,$$

is $\le t$; its rank over $\mathbb{F}_q$ is $\le mt$. Hence the $\mathbb{F}_q$-dimension of the code is $\ge n - mt$.

The well-known equivalence of these two definitions, which itself is a simple consequence of the Lagrange interpolation formula, underlies much of the usefulness of the Goppa codes. Later, after we have realized the construction of the Goppa codes as embeddings of the projective line $\mathbb{P}^1$ into higher-dimensional projective spaces, we will see that the above equivalence is just a simple manifestation of the (algebraic–geometric) duality between rational functions and differentials on an algebraic curve. The recent generalization given by Goppa for his codes rests on the highly developed theory of algebraic curves over finite fields.

The simple structure of the parity check matrix, which makes the Goppa codes easy to encode, and the representation of code words by rational functions, which reduces the decoding problem to an equivalent one in diophantine approximations, are two of the most interesting properties of these class of codes which makes them quite practical for implementation. On the theoretical side special sequences of codes in this family possess excellent parameters.

For reference later we gather in the following statement the basic properties of the Goppa codes.

Theorem 5.1 *With the notation as above, the Goppa code $\Gamma(L, g)$ associated with the data $D = \{L, g\}$ has the following properties:*

- (i) $\dim_{\mathbb{F}_q} \Gamma(L, g) \ge n - mt$,
- (ii) *min. weight of $\Gamma(L, g) \ge t + 1$,*
- (iii) *$\Gamma(L, g)$ is easily decoded by a continued fraction algorithm,*
- (iv) *there is a sequence of Goppa codes over $\mathbb{F}_q$ which meet the Varshamov–Gilbert bound.*

Remark As we indicated above (i) and (ii) are easy consequences of the two definitions of $\Gamma(L, g)$. After the introduction of the language of algebraic curves over finite fields and their associated function fields, we shall consider the question of decoding in its equivalent formulation as a problem of best approximations; it will turn out that the basis of the decoding algorithm for Goppa codes, say as implemented in the well-known method of Berlekamp, Massey & Patterson [5], is the simple fact, first observed by Huyghens, that the best rational approximations to a real number are provided by the convergents of its continued fraction. As for (iv) we shall see later on, that the more general theory of Goppa codes constructed from algebraic curves goes beyond the limits delineated by the Varshamov–Gilbert bound.

As a first approximation to the general construction of codes on an algebraic curve we will now realize the Goppa codes defined above as arising from the geometry of the projective line.

5.2 The affine and projective lines

The most basic object of study in the theory of algebraic curves is the projective line. Among the various possible approaches to its study we shall follow the arithmetic one. This will permit a clear and constructive consideration of certain geometric questions relative to the field of constants; this is quite suitable to the construction of codes, where the fields of constants are usually finite extensions of the Galois field $\mathbb{F}_2$.

As a preliminary to the arithmetic study of the function field of the projective line and other related concepts we begin with the simpler concept of the affine line.

5.2.1 Affine line $\mathbb{A}^1(k)$

Over an algebraically closed field k, the points on the affine line $\mathbb{A}^1(k)$ are in one-to-one correspondence with the elements of the field k and these in turn can be made to correspond to the monic polynomials of degree 1 in one indeterminate with coefficients in k; by abuse of language we will refer to the family of polynomials $\{T - a: a \in k\}$ as the set of closed points on the affine line $\mathbb{A}^1(k)$. Since for our purposes the field k will not be algebraically closed, we need to consider a relative notion, in which the points on $\mathbb{A}^1(k)$ correspond to orbits of points under the action of the group of automorphisms of some algebraic closure of k. We are thus led to introduce the following notion.

Definition 5.3 Let $k = \mathbb{F}_q$ be a finite field. The monic irreducible polynomials in $k[T]$ of degree d in the indeterminate T will be called the closed points of the affine line $\mathbb{A}^1(k)$ rational over k and of degree d. If the point P corresponds to the polynomial $f(T)$ of degree d, we will write $\deg P = \deg f = d$.

Remark It should be observed that this is a relative notion which depends on the field of constants k.

Example The following is a complete list of the closed points on the affine line $\mathbb{A}^1(\mathbb{F}_2)$ of degree 4:

$$P_1 = T^4 + T + 1;$$

$$P_2 = T^4 + T^3 + 1;$$

$$P_3 = T^4 + T^3 + T^2 + T + 1.$$

5.2.2 Projective line $\mathbb{P}^1$

The projective line will be obtained by adding to the affine line $\mathbb{A}^1(k)$ a 'point at infinity': $\mathbb{P}^1(k) = \mathbb{A}^1(k) \cup \{\infty\}$. To make this idea into a precise and workable definition we need to consider the functions defined on $\mathbb{A}^1(k)$; these are the elements of the field of rational functions $K = k(T)$ in the transcendental element T. The finite closed points (as opposed to the point at infinity) defined above correspond to discrete valuation rings (DVR) of k. Recall that a *discrete valuation* of K is a mapping $v = \text{ord of } K^x$ onto $\mathbb{Z}$ (where $K^x = K - \{0\}$ is the multiplicative group of K) such that

(i) $v(xy) = v(x) + v(y)$, i.e. v is a homomorphism
(ii) $v(x + y) \geq \min(v(x), v(y))$.

The set consisting of 0 and all $x \in K^x$ such that $\text{ord}(x) \geq 0$ is the discrete valuation ring associated to v. It is often convenient to extend v to the whole of K by putting $v(0) = \infty$.

How is the above connected to the notion of point on the affine line? Since the ring of polynomials $k[T]$ is a unique factorization domain, any rational function $f(T)$ in $k(T)$ can be uniquely represented, up to a constant factor as a product of monic irreducible polynomials each appearing with a certain multiplicity. Therefore if $p(T)$ is a monic irreducible polynomial corresponding to the point P in $\mathbb{A}^1(k)$, we can define a valuation v_P by

$$\text{ord}_P : k(T)^x \to \mathbb{Z}$$

$$f \to \text{ord}_P(f) = e,$$

where f is represented as a quotient of polynomials

$$f = \frac{r(x)}{s(t)} p(T)^e,$$

where r, s and p are relatively prime polynomials. It is immediate that this defines a discrete valuation on the field $k(T)$ and hence with the point P we can associate the discrete valuation ring

$$R_P = \{f \in k(T): \text{ord}_P(f) \geq 0\}.$$

It will often be convenient to refer to this ring as the ring of rational functions on the affine line $\mathbb{A}^1(k)$ which are free of poles at the point P.

In the field $K = k(T)$ there is another discrete valuation which does not arise from those constructed above; it is associated to the rational function $p_\infty(T) = 1/T$; in the following we will refer to it as the 'polynomial corresponding to the point at infinity'. The discrete valuation corresponding to $p_\infty(T) = 1/T$ is defined by setting $\mathrm{ord}_\infty(f) = r$, where we have represented $f(T) = s(T)/r(T)$ as the quotient of 2 polynomials in $k[T]$ and $r = $ degree of $r(T) - $ degree of $s(T)$. For instance, if $s(T)$ is a polynomial of degree d, then $\mathrm{ord}_\infty s = -d$. It is easily verified that ord_∞ is indeed a discrete valuation and its associated ring is

$$R_\infty = \{ f \in k(T) : \mathrm{ord}_\infty(f) \geq 0 \}.$$

It will often be convenient to think of the point at infinity as a point rational over k; we shall in the following refer to this by saying that the point P_∞ has degree 1 over k.

Remark By choosing a real number c, $0 < c < 1$, a discrete valuation v of $k(T)$ makes the set $k(T)$ into a metric space by defining the distance between two elements $x, y \in k(T)$ to be

$$d(x, y) = c^{v(x-y)};$$

hence v induces a topology on $k(T)$. Two discrete valuations which induce the same topology are called equivalent. A more precise notion than that introduced above would be that of an equivalence class of discrete valuations; in accord with the constructive point of view followed in coding theory, we will be interested only on the (normalized) discrete valuations introduced above and in the following we shall simply refer to these as valuations.

The fundamental result about discrete valuations of the field $k(T)$ which we shall use repeatedly is the following.

Theorem 5.2 With notations and assumptions as above the only discrete valuations of $k(T)$ are (for a proof see [11])

  (i) $v_p = \mathrm{ord}_P$, where P is a monic irreducible polynomial in $k[T]$;
  (ii) $v_\infty = \mathrm{ord}_{P_\infty}$, where P_∞ is the rational function $P_\infty = 1/T$.

This theorem justifies the following definition.

Definition 5.4 Let k be a finite field. The closed points on the projective line $\mathbb{P}^1(k)$, rational over k, correspond in a one-to-one manner to the discrete *valuations of $k(T)$, the field of rational functions in the transcendental element T.*

Remark The role of the transcendental element T in the above definition is of an auxiliary nature. In fact the function field $k(T)$ has many other generators. From now on the algebraic geometry of the projective line $\mathbb{P}^1(k)$ will consist of the study of the discrete valuation rings of the function field $k(T)$, where T is a transcendental element.

In concrete situations where $k(T)$ will appear as the field of rational functions on the projective line $\mathbb{P}^1(k)$, the function $p_\infty(T) = 1/T$ will be called a local uniformizer parameter at the point P_∞ at infinity. Analogously, if the closed point P corresponds to the monic irreducible polynomial $p(T)$ in $k[T]$, then $p(T)$ will be called a local uniformizing parameter (l.u.p.) at P. If P is a closed point of $\mathbb{P}^1(k)$ and t is the corresponding l.u.p., then $\text{ord}_P(t) = 1$ and in particular, the local ring

$$R_P = \{f \in k(T): \text{ord}_P(f) \geq 0\}$$

has

$$m_P = \{f \in k(T): \text{ord}_P(f) > 0\}$$

as a unique maximal ideal. It is not hard to verify from the definitions that the quotient $\mathbb{F}_P = R_P/m_P$ is a finite field extension of the field of constants k of degree equal to $\deg P$, i.e. for any closed point P on $\mathbb{P}^1(k)$ we have

$$\dim_k(R_P/m_P) = \deg P.$$

Definition 5.5 *A divisor on the projective line $\mathbb{P}^1(k)$ is a formal sum of points with integral (positive or negative) coefficients; that is an expression of the form*

$$D = \sum_P m_P P,$$

where the sum is over all closed points of $\mathbb{P}^1(k)$, and the integers $m_P \in \mathbb{Z}$ are all, except for a finite number, equal to 0.

A divisor is called *finite* if it does not contain the point at infinity with a non-zero coefficient. A divisor $D = \sum_P m_P P$ is called *positive* if all the coefficients m_P are positive. If $D = \sum_P m_P P$ and $D' = \sum_P n_P P$ are two divisors, we say that D is greater than D', and denote this by writing

$$D \geq D',$$

if all the corresponding coefficients satisfy $m_P \geq n_P$. In particular if 0 denotes the divisor all of whose coefficients are zero, then a positive divisor satisfies $D \geq 0$. It is clear that the set of all divisors forms an additive group. Finite positive divisors are essentially the same as the ideals in the ring $k[T]$; to every such divisor $D = \sum_P m_P P$ we attach the polynomial $p_D(t) = t^n + a_1 t^{n-1} + \cdots + a_n$ which generates the ideal, i.e. whose irreducible factors correspond to the points in D with multiplicities respectively equal to their coefficients in D. As n can be thought of as the degree of D, we are led to make the following definition.

Definition 5.6 *Let $D = \sum_P m_P P$ be a divisor on $\mathbb{P}^1(k)$. The degree of D is the integer*

$$\deg D = \sum_P m_P \deg P.$$

Remark As already indicated, this notion of degree is a relative one which depends on the field of constants k.

Definition 5.7 *To a rational function f in $k(T)$ we associate a divisor*

$$(f) = \sum_P \operatorname{ord}_P(f)P.$$

As a rational function of T is a quotient of two polynomials each of which is a product of at most a finite number of irreducible polynomials, it follows that $\operatorname{ord}_P(f) = 0$ for all except a finite number of points P and hence the divisor (f) is well defined. A divisor D which is of the form $D = (f)$ will be called *principal*. Since the quotient $f = r(T)/s(T)$ of two relatively prime polynomials $r(T)$, $s(T)$ in $k[T]$ of respective degrees d and d' is also expressible in terms of the function $p_\infty(T) = 1/T$ in the form

$$f(T) = p(T)^{d-d'} u\left(\frac{1}{T}\right) \Big/ v\left(\frac{1}{T}\right),$$

where u, v are polynomials in $1/T$ with non-zero constant terms, we arrive at the following result.

Theorem 5.3 *If $f \in k(T)$ then $\deg(f) = 0$, i.e. the degree of a principal divisor is 0.*

It is a basic and fundamental result of the geometry of the projective line that associated to any divisor D on $\mathbb{P}^1(k)$ with $\deg D = 0$, there is a function f in $k(T)$, unique up to a constant factor, such that $D = (f)$. This is easily seen if D is a finite divisor, in which case it is of the form $D = M - N$, where

M, N are each finite positive divisors; to D we attach the function $f_D(T) = p_M(T)/p_N(T)$; clearly $\deg D = 0$ implies $\deg p_M(T) = \deg p_N(T)$, and $f_D(T)$ is the unique function with $(f_D) = D$ and taking the value 1 at infinity.

Let f be a rational function in $k(T)$. If P is a point on the projective line $\mathbb{P}^1(k)$ and $\mathrm{ord}_P(f) > 0$, then P is called a *zero* of f; if $\mathrm{ord}_P(f) < 0$, then P is called a *pole* of f. Clearly the divisor of f can be thought of as describing the set of zeros and poles with their appropriate multiplicities. The value at P can be defined as follows: If $P = P_\infty$, the point at infinity, which is rational over k, we represent f as a formal power series in $p_\infty(T) = T^{-1}$

$$ f = a_0 + a_1\left(\frac{1}{T}\right) + a_2\left(\frac{1}{T}\right)^2 + \cdots, \qquad a_i \in k, $$

and we put $f(P_\infty) = a_0$. If $P \neq P_\infty$ corresponds to a monic irreducible polynomial $p(T)$ and f is represented in the form

$$ f = a \cdot r(T)/s(T), \qquad r, s \text{ monic}, $$

then by the Euclidean algorithm we can represent

$$ r(T) = p(T)q(T) + u(T), \qquad \deg u < \deg p $$

$$ s(T) = p(T)q_0(T) + u_0(T), \qquad \deg u_0 < \deg p. $$

Under the mapping 'reduction modulo $p(T)$':

$$ k[T] \to k[T]/(p) \simeq k_P, $$

the quotient $a \cdot u(T)/u_0(T)$ is represented by a well-defined element in k_P which we denote by $f(P)$; this we take as the *value* of f at P. We note that the value of a rational function f at a point P in general will not be an element in k unless P itself is of degree 1.

If P is a closed point and f is a function with at most one pole at P of multiplicity m_P, then it follows from the definitions that $(f) + m_P P \geq 0$. More generally if $D = D_0 - D_\infty$ is the difference of two disjoint (no points in common!) positive divisors, i.e.

$$ D_0 = \sum_P m_P P, \qquad m_P \geq 0 $$

$$ D_\infty = \sum_Q n_Q Q \qquad n_Q \geq 0 $$

and $\min(m_P, n_P) = 0$, then the relation $(f) + D \geq 0$ specifies that f has poles at the points P in D_0 with multiplicities bounded by m_P and zeros at P with multiplicities at least n_P. Conversely, given a divisor D we can ask for the set of functions $f \in k(T)$ which satisfy the relation $(f) + D \geq 0$; the study

of such a set is one of our main goals. As a consequence of the definitions
it is readily verified that this set together with the function identically 0
forms a vector space over the field of constants. This motivates the following definition.

Definition 5.8 *Let D be a divisor on* $\mathbb{P}^1(k)$. *The linear space associated with
D is*

$$L(D) = \{f \in k(T): (f) + D \geq 0\} \cup \{0\}.$$

We also put $l(D) = \dim_k L(D)$.

The following is the most primitive version of the Riemann–Roch theorem
applied to $\mathbb{P}^1(k)$.

Theorem 5.4 *We have* $l(D) = \deg D + 1$.

In the present context the result is an easy consequence of the theorem
on the independence of the discrete valuations on the field $k(T)$ arising
from the distinct monic irreducible polynomials of $k[T]$ and the function
$p_\infty(T) = 1/T$. After we unwind the definitions and make the relation $(f) +
D \geq 0$ explicit in terms of local uniformizing parameters the proof becomes
a problem of linear algebra. For instance if t is a positive integer $L(tP_\infty)$
is simply the vector subspace of polynomials in $k[T]$ of degree $\leq t$; the
k-dimension of this subspace is clearly $t + 1$. The general case follows by
induction on the degree of D and a comparison of $L(D)$ and $L(D + P)$. We
shall not give the details here since later we shall present the full statement
of the Riemann–Roch Theorem.

For the purpose of reinterpreting the definitions of Goppa codes as given
in Section 5.1 in terms of the projective line, we give now a geometric
description which is quite suitable when the field of constant k is algebrai-
cally closed, say $k = \mathbb{C}$ the complex numbers or $k = \bigcup_{d=1}^{\infty} \mathbb{F}_{q^d}$, the algebraic
closure of $\mathbb{F}_q$. In the affine plane $\mathbb{A}^2 = k^2$, where the points correspond to
pairs of elements (x_1, x_2), called coordinates, one considers the set of lines
through the origin $(0, 0)$; the mapping

$$L \to \text{slope } L,$$

which associates the line $x_2 = \lambda x_1$ to its slope λ, gives a correspondence
between the set of lines $\{L\}$ and the elements in $k \cup \{\infty\} = \mathbb{P}^1(k)$. Thus the
points on $\mathbb{P}^1(k)$ can be identified with the equivalence classes of points in
$k^2 - (0, 0)$, where $(x_0, x_1) \sim (y_0, y_1)$ if $x_0 = \lambda y_0$ and $x_1 = \lambda y_1$ for some
$\lambda \in k^x$. In this realization the point at infinity P_∞ is represented by the point
with coordinates $(1, 0)$.

Example The set $P_\infty = (1,0)$, $P_\alpha = (\alpha, 1)$, $\alpha \in \mathbb{F}_q$, represents all points on $\mathbb{P}^1(\mathbb{F}_q)$ of degree 1 over $\mathbb{F}_q$.

As above, when k is algebraically closed, one defines the points on projective n-space $\mathbb{P}^n(k)$ to consist of equivalence classes of $n + 1$ tuples $(x_0, x_1, \ldots, x_n)$ in k^{n+1} with at least one $x_i \neq 0$, where equivalence is defined by $(x_0, x_1, \ldots, x_n) \sim (y_0, y_1, \ldots, y_n)$ if $x_i = \lambda y_i$, $0 \leq i \leq n$, for some $\lambda \in k^x$. For example the projective plane consists of all lines through the origin in k^3 plus the point at infinity.

In Goppa's conception of the relation between algebraic geometry and coding theory, the study of the mappings

$$f: \mathbb{P}^r \to \mathbb{P}^s$$

is of great interest. Here $f(P) = (f_0(P), f_1(P), \ldots, f_s(P))$ and the f_i are homogeneous forms in the coordinates of the point P and of the same degree. It is not true that all such maps are well defined, since it may occur that $f_i(P) = 0$ for all i. Among those cases where the map is indeed defined, the following example is of particular interest.

Example Let d be a positive integer and let $f_0, f_1, \ldots, f_N$ be a basis for the vector space of homogeneous forms of degree d in x_0, x_1. The mapping

$$\mathbb{P}^1(k) \to \mathbb{P}^N(k)$$

$$P \to (f_0(P), f_1(P), \ldots, f_N(P))$$

embeds the projective line into $\mathbb{P}^N(k)$ where it now beomes a curve of degree d.

In Section 5.5 we shall return to these mappings in the study of linear series on algebraic curves.

5.3 Goppa codes on the projective line

We now return to the definition of the codes of Section 5.1 and give a comparison from the point of view of constructions on the projective line. Recall that the relevant data consists of a (monic) polynomial $g(T)$ of degree t with coefficients in $\mathbb{F}_{q^m}$, and of a set of n distinct elements $L = \{\alpha_1, \ldots, \alpha_n\}$ in $\mathbb{F}_{q^m}$ different from the zeros of $g(T)$. In Definition 5.1 one is interested in those code words $c = (c_1, \ldots, c_n)$ in $\mathbb{F}_q^n$ for which the relation

$$f(T) = \sum_{i=1}^{n} \frac{c_i}{T - \alpha_i} \equiv 0 \bmod g(T) \tag{5.2}$$

holds. These functions form a subspace of the vector space of all functions $f(T)$ which satisfy the same condition but where the coefficients c_i are allowed to lie in $\mathbb{F}_{q^m}$. As a first approximation we can describe this latter faimly of functions as those rational functions on the projective line $\mathbb{P}^1(\mathbb{F}_{q^m})$ which have poles at the 'points' $T - \alpha_i$, $1 \leq i \leq n$, each of multiplicity at most one, and which vanish at the 'point at infinity' as well as at the zeros of $g(T)$. To make this remark precise we use the notation introduced in Section 5.2. First we consider the divisor (rational over $\mathbb{F}_{q^m}$)

$$D = P_1 + \cdots + P_n - P_\infty - \sum_P m_P P,$$

where $P_1, \ldots, P_n$, P_∞ are the closed points on $\mathbb{P}^1(\mathbb{F}_{q^m})$ which correspond to the polynomials $p_i(T) = T - \alpha_i$, $1 \leq i \leq n$, and $p_\infty(T) = 1/T$. The positive divisor $G = \sum_P m_P P$ corresponds to the factorization

$$g(T) = \prod_P p_P(T)^{m_P}$$

into a product of monic polynomials, irreducible over $\mathbb{F}_{q^m}$, each appearing with multiplicity m_P. The condition (5.2) is equivalent to the condition

$$(f) + D \geq 0 \qquad (5.3)$$

on the principal divisor generated by $f(T)$. As remarked in Section 5.2, the functions which satisfy (5.3) with zero adjoined form a vector space $L(D)$ over $\mathbb{F}_{q^m}$ of dimension

$$l(D) = \deg D + 1$$

$$= n - \deg g,$$

and hence $\dim_{\mathbb{F}_q} L(D) = m(n - \deg g)$. In defining the Goppa code $\Gamma(L, g)$ we consider only code words $c = (c_1, \ldots, c_n) \in \mathbb{F}_q^n$ for which the corresponding rational function

$$f(T) = \sum_{i=1}^n \frac{c_i}{T - \alpha_i}$$

belongs to $L(D)$. We have already obtained in Section 5.1 the lower bound

$$\dim_{\mathbb{F}_q} \Gamma(L, g) \geq n - mt.$$

When $m = 1$ we actually have equality; a basis for $L(D)$ in this case is easily seen to be

$$g(T)T^i/f(T), \qquad 0 \leq i \leq n - t - 1,$$

where $f(T) = (T - \alpha_1) \ldots (T - \alpha_n)$. There are other cases where the dimension of $\Gamma(L, g)$ can be calculated precisely. For example if all the $\alpha_i \in \mathbb{F}_q$, then the code word $c = (c_1, \ldots, c_n) \in \mathbb{F}_q^n$ is associated with the rational function

$$r(T) = \sum_{i=1}^{n} \frac{c_i}{T - \alpha_i} = \frac{A(T)}{B(T)}$$

in $\mathbb{F}_q(T)$ with $A(T) = g(T)h(T)$; now if σ is an automorphism in $\mathrm{Gal}(\mathbb{F}_{q^m}/\mathbb{F}_q)$ then $A(T) = g(T)^\sigma h(T)^\sigma$ and hence $A(T)$ is divisible by all the conjugates of the polynomial $g(T)$. Suppose the number of distinct conjugates is s. Then $\dim_{\mathbb{F}_q} \Gamma(L, g) = n - st$.

Remark In order to facilitate the calculation of the dimension of $\Gamma(L, g)$ in the case where the projective line is replaced by an algebraic curve, in the following we shall restrict our attention to the case $\mathbb{F}_{q^m} = \mathbb{F}_q$ where we can use the Riemann–Roch theorem directly.

Before we pass to the second definition of $\Gamma(L, g)$ we should point out that the main benefit to be gained from the above rational representation lies in the control which we have over the weight of a code word. Let $w = w(c)$ be the weight of $c = (c_1, \ldots, c_n)$ and suppose the poles of the corresponding rational function $r(T)$ are in the set $\{Q_1, \ldots, Q_w\} \subset \{P_1, \ldots, P_n\}$. Then the vector space

$$L\left(Q_1 + \cdots + Q_w - P_\infty - \sum_P m_p P \right)$$

is not empty and hence its dimension satisfies

$$w - \deg g \geq 1,$$

that is $w \geq 1 + \deg g$. Hence the weight of $\Gamma(L, g)$ is $\geq 1 + \deg g$. This aspect of the rational representation of $\Gamma(L, g)$ generalizes without difficulty to codes over algebraic curves.

Let us consider a closed point P on the projective line $\mathbb{P}^1(\mathbb{F}_q)$, of degree d. For simplicity of notation we denote the polynomial corresponding to P by

$$t_P = p_P(T),$$

or simply by t if no confusion arises. Let $r(T) \in \mathbb{F}_q(T)$ and suppose that in terms of the local uniformizing parameter t_P it is representable as a formal power series by

$$r(T) = \sum_{\mu=v}^{\infty} a_\mu t_P^\mu, \qquad a_\mu \in \mathbb{F}_{q^d}; \tag{5.4}$$

the formal derivative of r is

$$r'(T) = \sum_{\substack{\mu=v \\ \mu \neq 0}} \mu a_\mu t_P^{\mu-1}.$$

This applies in particular to

$$T = \sum_{\mu=v}^{\infty} b_\mu t_P^\mu$$

$$dT = \sum_\mu \mu b_\mu t_P^{\mu-1} \, dt_P.$$

(5.5)

Remark A precise definition of the symbols dT, dt_P will be given in Section 5.4. Thus to any formal expression of the type

$$\omega = r(T) \, dT$$

there is associated a divisor, called the *divisor* of the differential and denoted by (ω). First we define

$$\text{ord}_P(\omega)$$

as the smallest exponent e in the formal power series of ω in powers of t_P, i.e.

$$r(T) \, dT = \sum_{\mu=e}^{\infty} a_\mu t_P^\mu \, dt_P, \qquad a_e \neq 0.$$

If we think of T as the local uniformizing parameter of a point Q in $\mathbb{P}^1$, then clearly

$$dT = dt_Q \quad \text{and} \quad \text{ord}_Q(dT) = 0.$$

On the other hand, $p_\infty(T) = 1/T = t_\infty$ is the local uniformizing parameter for the point P_∞ at infinity and in this case $dT = -t_\infty^{-2} \, dt_\infty$, hence $\text{ord}_\infty(dT) = -2$. As the function T on $\mathbb{P}^1$ has no other poles or zeros, the divisor of dT is $(dT) = -2P_\infty$. The fact that a function has at most a finite number of poles and zeros implies the divisor

$$(\omega) = \sum_P \text{ord}_P(\omega) P$$

is well defined.

Let P be a closed point of degree 1 and let

$$\omega = \sum_\mu a_\mu t_P^\mu \, dt_P$$

be the formal power series expansion of the differential ω in terms of the local uniformizing parameter t_P. The *residue* of ω at P is defined by

$$\text{res}_P(\omega) = a_{-1}.$$

(5.6)

Observe that this is an element of $\mathbb{F}_q$.

If the function f is an element in $L(D)$, i.e. $(f) + D \geq 0$, then the differential $\omega = f\,dT$ satisfies $(\omega) + D - 2P_\infty \geq 0$. Denote by $\Omega(P_1 + \cdots + P_n + P_\infty - G)$ the vector space of differentials which have poles at the points $P_1, \ldots, P_n$, P_∞ with multiplicity at most 1 and zeros on the support of the divisor $G = \sum_P m_P P$ with multiplicities at least m_P, i.e.

$$(\omega) + P_1 + \cdots + P_n + P_\infty - G \geq 0.$$

Multiplication of a function $f(T)$ by dT increases the order of the pole at infinity by at most 2 and introduces no other pole. Since a function f in $L(P_1 + \cdots + P_n - P_\infty - G)$ has a zero at infinity, $f(T)\,dT$ has at most a simple pole. Hence the mapping $f \to f \cdot dT$ provides an isomorphism of vector spaces

$$L(P_1 + \cdots + P_n - P_\infty - G) \simeq \Omega(P_1 + \cdots + P_n + P_\infty - G).$$

Using the notation introduced above we can redefine the Goppa code $\Gamma(L, g)$ as the image of the residue map

$$\text{res}: \Omega(P_1 + \cdots + P_n + P_\infty - G) \to \mathbb{F}_q^n$$

$$\omega \to (\text{res}_{P_1}\omega, \ldots, \text{res}_{P_n}\omega).$$

Since there are no non-constant functions on $\mathbb{P}^1(\mathbb{F}_q)$ free of poles, the mapping 'res' is one-to-one and hence the dimension of the image is $n - \deg G$, i.e. the dimension of $L(P_1 + \cdots + P_n - P_\infty - G)$.

Definition 2 of the Goppa code $\Gamma(L, g)$ can now be stated as follows. Consider the vector space

$$L(G - P_\infty) = \{f \in \mathbb{F}_q(T): (f) + G - P_\infty \geq 0\}$$

which is of dimension $\deg(G - P_\infty) + 1 = \deg G = t$, the degree of g; let $f_1, \ldots, f_t$ be a basis. The parity check matrix of $\Gamma(L, g)$ is given by

$$H = (f_i(P_j)), \qquad 1 \leq i \leq t$$

$$1 \leq j \leq n.$$

To establish the equivalence of the two definition of $\Gamma(L, g)$ we now prove the following lemma.

Lemma 5.1 *Let $c = (c_1, \ldots, c_n)$ be a code word in $\Gamma(L, g)$ and let f be an element in $L(G - P_\infty)$. Then we have*

$$\sum_{j=1}^{n} f(P_j)c_j = 0.$$

Remark The following proof is given here because it exemplifies in a trivial case the *residue theorem* which is the second fundamental result in the theory of algebraic surves used in the construction of Goppa codes given in Section 5.5.

Proof. Observe first that the differential

$$r(T) \cdot dT = \sum_{i=1}^{n} \frac{c_i \, dT}{T - \alpha_i}$$

has at worst a simple pole at infinity, hence, since f vanishes at P_∞, the differential

$$f(T)r(T)\,dT$$

does not have a pole at P_∞ and in particular its residue there is 0. The residue of the differential

$$\frac{c_i \, dT}{T - \alpha_i} \tag{5.7}$$

at the point P_i corresponding to the polynomial $p_i(T) = T - \alpha_i$ is clearly c_i. Now the expression (5.7) has also a pole at P_∞ and its residue is $-c_i$. From the identity

$$\frac{c_i f(T)\,dT}{T - \alpha_i} = \frac{c_i f(\alpha_i)\,dT}{T - \alpha_i} + \frac{c_i(f(T) - f(\alpha_i))\,dT}{T - \alpha_i}$$

and the fact that $f(1/t_\infty)$ is a power series in t_∞ without constant term, it follows that

$$\mathrm{res}_\infty\left(\frac{f(T)c_i\,dT}{T - \alpha_i}\right) = -f(P_i)c_i.$$

Since the residue map is additive and the differential $\omega = f(T)r(T)\,dT$ has no pole at P_∞ it follows that the sum of the residues at P_∞ is 0, i.e.

$$\sum_{i=1}^{n} c_i f(P_i) = 0.$$

This proves the lemma.

As a final note to this section and as motivation for the study of algebraic curves other than the projective line in connection with coding theory we make the following two remarks.

Remark An important property of the projective line which we have not yet used but which plays a role in the proof of the existence of a sequence of Goppa codes which meet the Varshamov–Gilbert bound is that on $\mathbb{P}^1(\mathbb{F}_{q^m})$ the number of points of degree t is quite large, roughly of the order of magnitude $(1/t)q^{mt}$ (see Van Lint [91], p. 111). The most significant advance in the theory of Goppa codes has come from the realization that on algebraic curves with a high degree of complexity, say as measured by the genus, and with sufficiently many rational points it is possible to construct codes, over certain fields $\mathbb{F}_q$, with rates that are better than the Varshamov–Gilbert bound.

Remark If the divisor G associated with $\Gamma(L, g)$ has degree t, then the vector space $L(G - P_\infty)$, of dimension t, gives rise to a projective space $\mathbb{P}^{t-1}$. Now a choice of a basis $f_0, \ldots, f_{t-1}$ gives rise to a mapping

$$\mathbb{P}^1 \to \mathbb{P}^{t-1}$$

$$P \to (f_0(P), \ldots, f_{t-1}(P)).$$

This realizes the projective line as a rational curve C of degree t in $\mathbb{P}^{t-1}$. The columns of the parity check matrix for $\Gamma(L, g)$ can be thought of as points on C. Since a curve of degree $t - 1$ meets a straight line in at most $t - 1$ points, it follows that any t points cannot lie on the same line; hence the weight of the corresponding code $\Gamma(L, g)$ is $t + 1$.

5.4 Algebraic curves

In describing the basic theory of algebraic curves over finite fields we shall rely on the arithmetic model of the projective line as presented in Section 5.2. To motivate the development we start with some general remarks on what is meant by the birational geometry on an algebraic curve. Let us quote from Lefschetz' illuminating article [50]:

Let k be a field of complex numbers, and let $f \in k[x, y]$ be an irreducible polynomial, and consider the curve

$$C: f(x, y) = 0.$$

Any complex rational function

$$R(x, y) = P(x, y)/Q(x, y),$$

where P and Q are polynomials, is declared to be zero: $R(x, y) = 0$, whenever P but not Q is divisible by f. Place in one class S^* all the rational functions which differ by zero from a given one. The collection $\{S^*\}$ is a field, which is an extension of the field k of all complex numbers. This is the function field $K = k(C)$ of the curve C. The algebraic geometry of C is the study of the properties of C which depend solely upon the function field K.

For applications to number theory and coding theory we consider a finite field of constants $k = \mathbb{F}_q$. We let f be a polynomial in $k[x, y]$ which is absolutely irreducible, i.e. which remains irreducible over any finite extension of k. This condition ascertains that the curve

$$C: f(x, y) = 0$$

is connected. Denote by (f) the principal ideal in $R[x, y]$ generated by f. The *coordinate ring* of the curve C is defined to be the quotient

$$R = k[x, y]/(f).$$

The quotient field of R is the function field $K = k(C)$ of the curve C. It is an extension of the field of constants k of transcendence degree 1.

5.4.1 Separable extensions

Let k be a field. A polynomial in $k[X]$ is called *separable* if all its irreducible factors have distinct roots in some splitting field extension. For example if the characteristic of k is 0, then every polynomial in $k[X]$ is separable. It is easy to show that if $k = \mathbb{F}_p(t)$, t an indeterminate, then the polynomial $f = X^p - t$ is irreducible in $k[X]$, nevertheless $f = T^p - t$ is a p-th power in $E_f[T]$, where E_f is a splitting field extension for f. A field k (of any characteristic) is called *perfect* if every polynomial in $k[X]$ is separable. Clearly any field of characteristic 0 is perfect. It is well known that any field of positive characteristic p is perfect if and only if $k = k^p$, the subfield of pth powers of the elements of k. In particular, every finite field $\mathbb{F}_q$ is perfect. An extension field K of k is said to be algebraic over k if every element of K is algebraic over k; this is certainly the case if, as a vector space, K is of finite dimension over k. An extension K/k is called separable (algebraic) if the minimum polynomial of every element of K is separable. If K is algebraic over k, then K is necessarily separable over k if $\operatorname{char}(k) = 0$ or if $\operatorname{char}(k) = p$ and $k^p = k$. An extension K of a field k is said to be *separably generated* over k if there exists an intermediary field H between k and K which is purely transcendental over k and over which K is algebraic and separable. A field K of algebraic functions of one variable, i.e. the function $K = k(C)$ of an algebraic curve C, is said to be *separably generated* if it is separably generated over its field of constants; when this is the case, any element $x \in K$ such that K is algebraic and separable over $k(x)$ is called a *separating variable* in K. The following well-known result (see Eichler [24], p. 143) is not difficult to prove.

Proposition 5.1 *The function field $K = K(C)$ of an algebraic curve with a perfect field of constants k is separably generated.*

Remark This result applies in particular to the case $k = \mathbb{F}_q$ of main interest in coding theory.

5.4.2 Closed points and their neighborhoods

As indicated in Section 5.2, a discrete valuation on a field K is an onto homomorphism $v\colon K^\times \to \mathbb{Z}$ satisfying (i) $v(xy) = v(x) + v(y)$, (ii) $v(x + y) \geq \min(v(x), v(y))$. The set consisting of 0 and all the elements $x \in K^\times$ with $v(x) \geq 0$ forms a ring R_v, which we call a discrete valuation ring. It is often convenient to extend the definition of v to all of K by setting $v(0) = \infty$. The subset $m_v = \{x \in R_v\colon v(x) > 0\}$ is a maximal ideal of R_v. R_v and m_v are uniquely determined by the equivalence class of the valuation v, here equivalence $v \sim v'$ means that the topologies on K as defined by the metrics $d(x, y) = c^{-v(x-y)}, d'(x, y) = c^{-v'(x-y)}$, some real number c with $0 < c < 1$, are equivalent. It is easy to prove from the definitions that m_v is the only maximal ideal of R_v; such a ring R_v with a unique maximal ideal is called a local ring. The quotient $k_v = R_v/m_v$ is called the *residue class field* of the local ring (R_v, m_v). If $K = k(C)$ is the function field of an irreducible curve C and v is a discrete valuation of K, then the following are equivalent properties of the local ring R_v (see Atiyah & Macdonald [2], p. 94):

  (i) R_v is a discrete valuation ring;
  (ii) R_v is integrally closed, i.e. any element of the field of quotients which is integral over R_v is already an element of R_v;
  (iii) m_v is a principal ideal;
  (iv) $\dim_{k_v}(m_v/m_v^2) = 1$;
  (v) every non-zero ideal is a power of m_v;
  (vi) there exists an element $t \in R_v$ such that every non-zero ideal is of the form (t^a), $a \geq 0$.

Since we are not assuming that the field of constants k is algebraically closed, the residue class field of k_v is a finite extension of k. The degree of this extension is called the *degree of the valuation*. The element t of property (vi) is called a *local uniformizing parameter* (l.u.p.).

In the following definition we consider an irreducible algebraic curve C defined over the field k.

Definition 5.9 *A discrete valuation ring (R_v, m_v) of the function field $K = k(C)$ is called a closed point of C.*

When convenient we shall use the notation $P_v = (R_v, m_v)$ to denote the discrete valuation ring and the maximal ideal associated to the closed point P_v. If $k_v = R_v/m_v$ denotes the residue class field of P_v then we define

$$\deg(P_v) = [k_v : k]$$

to be the degree of the point P_v. Observe that the closed points are all of degree 1 only when k is algebraically closed.

Example If $K = k(T)$ is the function field of the projective line $\mathbb{P}^1(k)$ then a complete list of all the closed points is given by (i) those corresponding to monic irreducible polynomials $p(T)$ in $k[T]$, the so-called finite close points and (ii) the point P_∞ at infinity corresponding to the function $p_\infty(T) = 1/T$.

On a smooth irreducible algebraic curve $C: f(x, y) = 0$ defined over an algebraically closed field k, the notion of a closed point $P_v = (R_v, m_v)$ on C can be made to coincide with the notion of point $P = (\alpha, \beta)$ on C as a pair of elements α, β in k which satisfy $f(\alpha, \beta) = 0$. Let us briefly indicate how one passes from one notion to the other. Recall that the function field $K = k(C)$ is the field of quotients of the coordinate ring $k[x, y]/(f)$; as such any element g in K can be represented as quotient $g = A(x, y)/B(x, y)$ with $A, B \in k[x, y]$. With the point $P = (\alpha, \beta)$ we associate the rings

$$R = \left\{ g \in K : g = \frac{A(x, y)}{B(x, y)}, B(\alpha, \beta) = 0 \right\}$$

$$m = \left\{ g \in R : g = \frac{A(x, y)}{B(x, y)}, A(\alpha, \beta) = 0 \right\}.$$

If (α, β) is a simple point on $C: f(x, y) = 0$, i.e. $f_x(\alpha, \beta)$ and $f_y(\alpha, \beta)$ are not both zero, then the pair (R, m) is a discrete valuation ring and $P_v = (R, m)$ is the closed point corresponding to $P = (\alpha, \beta)$. Conversely, if $P_v = (R_v, m_v)$ is a closed point (of degree one), the maximal ideal m_v as an ideal in $k[x, y]/(f)$ must be of the form $m_v = (x - \alpha, y - \beta)$, with $\alpha, \beta \in k$, satisfying $f(\alpha, \beta)$.

When $P = (\alpha, \beta)$ is not a simple point on the curve $C: f(x, y) = 0$, there may be in general several closed points on $K = k(C)$ corresponding to P. In a naive sense this is related to the fact that a non-simple point may lie on several branches of the curve.

Example Let $C: f = 0$, where $f = y^2 - y - x^3 + x^2$. To test for simple points we consider the two equations

$$f_x = -3x^2 + 2x, \qquad f_y = 2y - 1.$$

Over the field of real numbers $\mathbb{R}$, f_x and f_y cannot be simultaneously equal to zero. Figure 5.1 gives part of the graph of the curve $y^2 - y - x^3 + x^2 = 0$

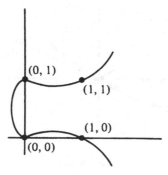

Figure 5.1

containing the closed points of degree one in the finite part of the plane. Over $k = \mathbb{F}_2, f_y \neq 0$ and hence all points are simple. In fact the closed points of degree 1 on $\mathbb{F}_2(C)$ are

$$P_1 = (0,0), \qquad P_2 = (0,1), \qquad P_3 = (1,0), \qquad P_4(1,1), \qquad P_\infty = (\infty, \infty);$$

here P_∞ stands for the point at infinity. These five points are the only points on C which are rational over $\mathbb{F}_2$. In general it is an easy consequence of the theory of the zeta function of C that the number N_d of closed points on $K = \mathbb{F}_{2^d}(C)$ of degree 1 is

$$N_d = 2^d + 1 - (-1 + i)^d - (-1 - i)^d, \qquad i = (-1)^{1/2}$$

Over $\mathbb{F}_3$ the points are also all simple. Let us now pursue the same example over a field of positive characteristic $p \neq 2, 3$. The only points which may fail to be simple are $P = (0, \frac{1}{2})$ and $Q = (\frac{2}{3}, \frac{1}{2})$; clearly P is not a point on the curve. It is a pleasant exercise to verify that the only prime p for which the equation

$$(\tfrac{1}{2})^2 - \tfrac{1}{2} - (\tfrac{2}{3})^3 + (\tfrac{2}{3})^2 \equiv 0 \bmod p$$

holds is $p = 11$. Hence the point $Q \equiv (8,6)$ is a non-simple point. Making the substitution $X = x - 8$, $Y = y - 6$, results in the equation of the curve C over $\mathbb{F}_{11}$ becoming

$$C: Y^2 = X^3 + X^2;$$

one checks easily that the origin $(0,0)$ is a double point with distinct tangents. Recall that over the field of real numbers $\mathbb{R}$, the curve $Y^2 = X^3 + X^2$ has the graph shown in Fig. 5.2. The mapping which sends a point P to the slope λ of the line through P and the origin shows that the function field $\mathbb{F}_{11}(C)$ is isomorphic to $\mathbb{F}_{11}(\lambda)$, i.e. the function field of the projective line $\mathbb{P}^1(\mathbb{F}_{11})$; the isomorphism is explicitly given by $X = \lambda^2 - 1$, $Y = \lambda(\lambda^2 - 1)$.

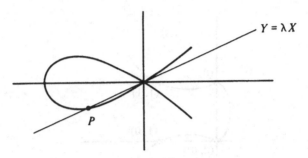

Figure 5.2

Example (the Fermat cubic) $C: u^3 + v^3 = w^3$ The substitution

$$x = \frac{3w}{u+v}, \qquad y = \frac{9}{2}\left(\frac{u-v}{u+v}\right) + \tfrac{1}{2}$$

transforms the curve into $C: f(x,y)$ with $f = y^2 - y - x^3 + 7$. The partial derivatives of f are $f_y = 2y - 1$, $f_x = -3x^2$. Over the field $\mathbb{F}_2$, $f_y \neq 0$; over a field of positive characteristic $p \neq 2$, $(0, \tfrac{1}{2})$ is the only possible non-simple point. In fact it is an easy exercise to show that the only prime p for which the congruence

$$(\tfrac{1}{2})^2 - \tfrac{1}{2} \equiv -7 \bmod p$$

holds is $p = 3$; hence $(0, 2)$ is a non-simple point on C over $\mathbb{F}_3$. The substitution $X = x$ and $Y = y - 2$ transforms the equation for C into

$$C: Y^2 \equiv X^3;$$

this model has a double tangent at the origin. Figure 5.3 exhibits part of the graph of C over the reals $\mathbb{R}$. A point such as $(0,0)$ is called a cusp. The mapping which sends a point $P \neq (0,0)$ on C to the slope λ of the line through P and $(0,0)$ sets an isomorphism between the function field $\mathbb{F}_3(C)$ and $\mathbb{F}_3(\lambda)$, the isomorphism being $X = \lambda$, $Y = \lambda^2$.

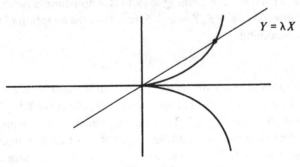

Figure 5.3

Table 5.1. *Points on* $y^2 - y = x^3 - 7$ *over* $\mathbb{F}_4 = \{0, 1, \alpha, \beta\}$.

	P_1	P_2	P_3	P_4	P_5	P_6	P_7	P_8	P_9
x	1	α	β	0	0	1	α	β	∞
y	1	1	1	α	β	0	0	0	∞

A well-known and very beautiful theorem of Gauss shows that for a prime $p \neq 3$, the number N of closed points on $C: y^2 - y = x^3 - 7$ of degree 1 over $\mathbb{F}_p$ is given by

$$p + 1 \qquad \text{if } p \equiv -1 \pmod 3,$$

$$p + 1 - a_p \qquad \text{if } p \equiv 1 \pmod 3,$$

where a_p is the unique integer $\equiv -1 \bmod 3$ such that $4p = a_p^2 + 27b_p^2$ for some integer b_p. For instance over $\mathbb{F}_2$ there are 3 closed points of degree 1. The theory of the zeta function of $\mathbb{F}_2(C)$ also shows that the number N_d of closed points on $\mathbb{F}_{2^d}(C)$ of degree 1 is

$$N_d = 2^d + 1 - \alpha^d - \beta^d, \qquad \alpha = \bar{\beta} = (-2)^{1/2}.$$

In particular, the 9 points on C over $\mathbb{F}_4$ are given by Table 5.1.

Let $k = \mathbb{F}_q$ and suppose $P_v = (R_v, m_v)$ is a closed point on an algebraic curve C. Let $k_v = R_v/m_v$ be the residue class field of P_v and let $t = t_v$ be the l.u.p. at P_v, i.e. a generator of m_v. By a process of successive approximations any element $f \in R_v$ satisfies

$$f \equiv \sum_{\mu=0}^{D} a_\mu t^\mu \bmod m_v^{D+1},$$

for any positive integer D, with unique elements $a_\mu \in k_v$. Passing to the limit which amounts to passing to the completion $\hat{R}_v$ of R_v with respect to the topology defined by m_v, we get a ring which is isomorphic to the ring of formal power series $k_v[[t]]$. The field of quotients of this ring is a field which we denote by K_v and which contains $K = k(C)$ as a dense subset. We restate this result in the case of interest here, namely when $k = \mathbb{F}_q$.

Proposition 5.2 *Let P_v be a closed point of degree d on $K = \mathbb{F}_q(C)$, and $t = t_v$ a local uniformizer parameter. The completion of K with respect to the discrete valuation v is isomorphic to $\mathbb{F}_{q^d}((t))$, the field of formal power series in t with coefficients in $\mathbb{F}_{q^d}$.*

Remark The above proposition plays a key role in the theory of algebraic curves since it allows us to think of a neighborhood of a simple point of degree d as looking like a portion of the projective line $\mathbb{P}^1(\mathbb{F}_{q^d})$, i.e. both have essentially the same function field.

Remark (Zariski topology) The curve C can be viewed as a topological space. Recall that a topology on a space consists of a set X endowed with a family of subsets T of X called the closed sets satisfying:

   (i) X and the empty set $\emptyset$ are closed,
   (ii) the union of any finite number of closed subsets is closed,
   (iii) the intersection of any number of closed subsets is closed.

The complement of a closed set is called *open*. With the set X of closed points of the function field $K = k(C)$ we associate a topology by decreeing that the closed sets are the subsets containing a finite number of closed points; the empty set and X itself are also considered as closed sets. In particular a closed point is also a *closed set*. In this setting it is appropriate to refer to the topological space X as the abstract curve associated with $k(C)$, and when there is no danger of confusion we write simply C for X. The abstract curve C as defined here is a complete topological space, which is in fact compact. Often it is also useful to consider abstract curves C where the underlying space X is missing a certain number of points. As usual, an open neighborhood of a point is an open set containing the point. Over the field of complex numbers $\mathbb{C}$ the abstract curve C is simply a Riemann surface.

Example The projective line $C = \mathbb{P}^1$ with function field $K = k(T)$ is an example of a complete curve. If P_∞ denotes the point at infinity, the complement $C - P_\infty$ is an open neighborhood, which as an abstract curve corresponds to the affine line $\mathbb{A}^1(k)$. Over an algebraically closed field k, the closed sets on $\mathbb{P}^1$ are defined by the vanishing of certain polynomials and as such are examples of the so-called algebraic sets, i.e. sets of points which are solutions of polynomial equations. This is a simple example of a Zariski topology.

5.4.3 Differentials

Let C be an irreducible curve defined over the field k with $K = k(C)$ as its function field. A k-linear mapping

$$D: K \to K$$

satisfying the Leibnitz rule $D(xy) = xDy + yDx$ for any two elements x, y of K is called a *derivation*. The set $\mathrm{Der}_k(K)$ of all derivations of K is a vector space over K. Its K-dual $\Omega_k(K)$ is the module of *differentials* on K; it is equipped with a k-linear mapping $d: K \to \Omega_k(K)$ which sends the element $x \in K$ to the differential dx. The duality between $\mathrm{Der}_k(K)$ and $\Omega_k(K)$ is given by $\langle D, dx \rangle = Dx$.

Remark In the following we shall use the notation Ω_C for the module $\Omega_k(K)$. It should be kept in mind that Ω_C is an object which intrinsically reflects the properties of the function field $K = k(C)$ and not those of C.

In general if A is a commutative algebra over k, i.e. a vector space over k in which a (commutative) product is defined, then a construction similar to that given above leads to the definition of the module $\Omega_k(A)$. Now let $x_1, \ldots, x_n \in A$ and $f \in k[x_1, \ldots, x_n]$ and define $f_i = \partial f/\partial x_i$ to be the formal derivative of f with respect to x_i. When $A = k[x_1, \ldots, x_n]$, the mapping

$$d: A \to \Omega_k(A)$$

given by

$$df = \sum_{i=1}^{n} f_i \, dx_i$$

shows that as an A-module, $\Omega_k(A)$ is generated by $dx_1, \ldots, dx_n$. In particular, after passing to quotients and extending d in the obvious manner $(d(f/g) = (g \, df - f \, dg)/g^2)$, we obtain that for $K = k(x, \ldots, x_n)$, the K-module $\Omega_k(K)$ is a finite-dimensional vector space over K generated by $dx_1, \ldots, dx_n$.

The module of differentials Ω_C has many pleasant properties which describe its behavior when the function field K is embedded in a larger field. For the moment it will be useful to recall the most fundamental property of Ω_C.

Proposition 5.3 *Let $K = k(C)$ be the function field of an algebraic curve C with field of constants $k = \mathbb{F}_q$. Then $\dim_K \Omega_C = 1$.*

Proof. Let $f \in k[X, Y]$ be a plane model for the curve C with function field K. Let $A = k[X, Y]/(f) = k[x, y]$ be the coordinate ring of C with K as its field of quotients. Since f is irreducible, i.e. $f_Y(x, y) \neq 0$, we may assume that f does not divide f_Y. As we saw above, dx and dy generate $\Omega_k(K)$. But $0 = d(f(x, y)) = f_X(x, y) \, dx + f_Y(x, y) \, dy$. Hence $dy = u \, dx$ with $u = -f_X(x, y)/f_Y(x, y)$. Therefore dx generates $\Omega_k(K)$ and $\dim_K \Omega_k(K) \leq 1$. It remains to show that $\Omega_k(K) \neq 0$. From the definition of the differential $d: K \to \Omega_k(K)$ it suffices to find a non-zero derivation $D: A \to V$, V some vector space over K. Let $V = K$ and for $g \in k[X, Y]$, let $\bar{g}$ be its image in the coordinate ring A, let $D(\bar{g}) = g_X(x, y) - u g_Y(x, y)$. It is easily verified that D is a well-defined derivation and $D(x) = 1$, so that $D \neq 0$. This proves the proposition.

Remark The above argument works in general for any algebraic curve whose field of constants k is perfect.

Let P_v be a closed point of degree d with residue class field k_v and l.u.p. $t = t_v$. If $f \in K = k(C)$ has a formal power series expansion

$$f = \sum_\mu a_\mu t^\mu, \qquad a_\mu \in k_v,$$

we have

$$df = \sum_\mu \mu a_\mu t^{\mu-1} dt,$$

where dt is the differential dual to the derivation D with $D(t) = 1$. In general any differential $\omega \in \Omega_C$ has a power series expansion about the point P_v of the form

$$\omega = \sum_\mu b_\mu t^{\mu-1} dt, \qquad b_\mu \in k_v.$$

Let m (resp. n) be the smallest exponent of a power of t which appears with a coefficient $a_m \neq 0$ (resp. $b_n \neq 0$) in the power series expansion of f (resp. ω).

Definition 5.10 *With notations as above, the orders of a function f and of a differential ω are given by*

$$\operatorname{ord}_v(f) = m, \qquad \operatorname{ord}_v^{\bullet}(\omega) = n.$$

If $m > 0$ (resp. $n > 0$) we say P_v is a *zero* of f (resp. of ω). If $m < 0$ (resp. $n < 0$) we say P_v is a *pole* of f (resp. of ω).

5.4.4 Divisors

By a divisor on C, rational over k, we mean a formal linear combination

$$D = \sum_P m_P P,$$

of closed points P with integral coefficients m_P all of which, except for a finite number, are zero. It is convenient to denote by 0 the zero divisor, i.e. the divisor with all $m_P = 0$. Clearly the set of all divisors form an additive group denoted by $\operatorname{Div}(C)$. If $D' = \sum_P n_P P$ is another divisor, we define the relation

$$D \geq D'$$

to mean that all the corresponding coefficients satisfy $m_P \geq n_P$. A divisor $D \geq 0$ is called a positive (sometimes effective) divisor.

Definition 5.11 *The degree of the divisor* $D = \sum_P m_P P$ *is*

$$\deg(D) = \sum_P m_P \deg(P).$$

This is a well-defined integer, since only a finite number of m_P are different from zero. The mapping deg: $\mathrm{Div}(C) \to \mathbb{Z}$ is clearly a homomorphism.

Since a function $f \in K(C)$ can have at most a finite number of poles or zeros, the following definitions make sense.

Definition 5.12 *The divisor of a function* $f \in K(C)$ *is*

$$(f) = \sum_P \mathrm{ord}_P(f)P.$$

The divisor of a differential $\omega \in \Omega_C$ *is*

$$(\omega) = \sum_P \mathrm{ord}_P(\omega)P.$$

The divisor (f) is often called the *principal divisor* associated to the function (f); its main property is given by the following proposition.

Proposition 5.4 *We have* $\deg((f)) = 0$ *for any function* $f \in K = k(C)$.

The proof of this result realizes K as a finite algebraic extension of the function field $k(x)$ of the projective line $\mathbb{P}^1$ together with the knowledge of how the closed points on C are related to the closed points on $\mathbb{P}^1$. The essential idea consists in reducing the assertion about f to the same assertion for the norm $N_{K/k(x)}(f)$ which is now a function on $\mathbb{P}^1$ where the claim is clear (see Section 5.2).

Since $\dim_K \Omega_C = 1$ and $(f\omega) = (f) + (\omega)$ for any $f \in K$ and $\omega \in \Omega_C$, the above proposition implies that the degree of a differential ω is a well-defined integer which depends only on K and not on ω. This suggests the following definition.

Definition 5.13 *The canonical class is the set of divisors*

$$W = \{(\omega) \in \mathrm{Div}(C) \colon \omega \in \Omega_C\}.$$

If $\omega \in \Omega_C$ *and* $\deg \omega = 2g - 2$, *we define the genus of* C *to be* g.

Equivalently we have number of zeros of ω – number of poles of $\omega = 2g - 2$.

It is not clear that g is positive or even an integer. This will follow from the proof of the Riemann–Roch theorem. What is clear from the definition is that g is a birational invariant of the curve, i.e. an object associated with the function field $K = k(C)$. For the projective line $C = \mathbb{P}^1$ we have already shown in Section 5.3 that any differential ω has degree -2, which implies that

$$\text{genus of } \mathbb{P}^1 = 0.$$

As we shall see later, the genus reflects to some extent the complex structure of a curve as compared to $\mathbb{P}^1$.

Definition 5.14 *Let P be a closed point on $K = \mathbb{F}_q(C)$ of degree d and let t be a local uniformizing parameter at P. If the differential $\omega \in \Omega_C$ has the formal power series expansion*

$$\omega = \sum_\mu a_\mu t^\mu \, dt, \qquad a_\mu \in \mathbb{F}_{q^d}.$$

Then the residue of ω at P is

$$\text{res}_P(\omega) = \text{tr}(a_{-1}),$$

where $\text{tr}(x) = x + x^q + \cdots + x^{q^{d-1}}$, for any $x \in \mathbb{F}_{q^d}$, is the trace mapping.

It is a pleasant exercise to verify that the above definition is independent of the l.u.p. chosen. It is also clear that $\text{res}_P \colon \Omega_C \to \mathbb{F}_q$ is an additive $\mathbb{F}_q$-linear mapping. For a closed point P of degree 1 we have $\text{res}_P(\omega) = a_{-1}$.

5.4.5 The theorems of Riemann–Roch, of Hurwitz and of the residue

As in Section 5.2 we let D be a divisor on C and define

$$L(D) = \{ f \in k(C)^* \colon (f) + D \geq 0 \} \cup \{0\}$$

If D is the difference of two positive divisors

$$D = \sum_P m_P \cdot P - \sum_Q n_Q \cdot Q,$$

Then the set $L(D)$ consists of all functions f in $k(C)$ such that the order of a pole at P cannot be larger than m_P, i.e. $\text{ord}_P(f) \geq -m_P$, and the order of a zero at Q has to be at least $-n_Q$, i.e. $\text{ord}_Q(f) \geq n_Q$. Since any discrete valuation v on $k(C)$ satisfies $v(\alpha f + \beta g) \geq \min(v(f), v(g))$ for any f, g in $k(C)$ and constants $\alpha, \beta \in k$, it follows that, after adjoining the zero element, $L(D)$ becomes a vector space. It is a consequence of the proof of the Riemann–

Roch theorem that $L(D)$ is of finite dimension over k, denoted in the following by

$$l(D) = \dim_k L(D).$$

As the number of zeros of a function is equal to the number of poles each counted with the appropriate multiplicity, it follows that if $\deg(D) < 0$ then $L(D)$ is empty and $l(D) = 0$. If two divisors D_1 and D_2 are in the same class, i.e. $D_1 - D_2 = (\varphi)$ a principal divisor, then multiplication by φ sets an isomorphism between $L(D_1)$ and $L(D_2)$; hence $l(D_1) = l(D_2)$ if $D_1 - D_2 = (\varphi)$.

In the construction of Goppa codes in Section 5.5 the following vector space of differentials will be useful:

$$\Omega_C(D) = \{\omega \in \Omega_C : (\omega) + D \geq 0\}.$$

Example On the projective line $\mathbb{P}^1$ with function field $K = \mathbb{F}_{q^m}(T)$ let P_i correspond to the polynomial $T - \alpha_i$, $1 \leq i \leq n$; Let $G = \sum_P m_P \cdot P$ be the (finite) divisor resulting from the factorization $g(T) = \prod_P p_P(t)^{m_P}$ into irreducible polynomials in $\mathbb{F}_{q^m}[T]$ (see Section 5.2); let also P_∞ be the point at infinity corresponding to $p_\infty(T) = 1/T$. In the construction of Goppa codes we have considered rational functions with poles at the points P_1, ..., P_n and divisible by the polynomial $g(T)$. If we let

$$D = P_1 + \cdots + P_n + P_\infty - G,$$

then, in terms of the notation introduced above, the expression

$$\omega = \sum_{i=1}^{n} \frac{c_i \, dT}{T - \alpha_i}$$

is precisely an element in the space of differentials $\Omega_{\mathbb{P}^1}(D)$.

Let ω be a fixed differential and (by abuse of notation) denote its divisor by $W = (\omega)$. Multiplication by ω induces an isomorphism of vector spaces

$$L(W + D) \rightarrow \Omega_C(D)$$

$$f \rightarrow f\omega.$$

In fact, a function $f \in L(W + D)$ satisfies $(f) + W + D \geq 0$; since $(f) + (\omega) = (f\omega)$, it is clear that $(f\omega) + D \geq 0$. The following lemma is now obvious.

Lemma 5.2 *We have* $\dim_k \Omega_C(D) = l(W + D)$.

We are now ready to state the two most important results in the arithmetic theory of algebraic curves.

Theorem 5.5 (Riemann–Roch) *Let D be a divisor on C and $W = (\omega)$ the divisor of a differential. Then we have*

$$l(D) = \deg D + 1 - g + l(W - D).$$

In particular if $\deg(W - D) = 2g - 2 - \deg D < 0$, then $l(D) = \deg D + 1 - g$.

Residue theorem 5.6 *Let $\omega \in \Omega_C$ be a differential. Then*

$$\sum_P \operatorname{res}_P(\omega) = 0,$$

where the sum $\sum_P$ is taken over all closed points P on C.

Remarks

1. There are presently available many proofs of the Riemann–Roch theorem. From the arithmetic point of view the most direct proof is the one due to Weil [101] which depends on the consideration of pseudo-differentials as duals of repartitions (pre-adeles). If the proof of the Riemann–Roch theorem is preceded by the proof of the residue theorem, then it is possible to extend Weil's proof to yield the statement given above. As for the residue theorem, since we are working over a finite field of constants, the best and simplest proof is that based on the *Cartier operator* ([24], p. 150).

2. In accordance with the constructive point of view of coding theory, it may be worth while to consider the older proof of the Riemann–Roch theorem due to Hansel and Landsberg, itself an outgrowth of the classic argument of Weber and Dedekind. The merit of this proof, which is more complex than that of Weil, is that it affords the construction of a basis for the space $L(D)$. When the field of constants is of characteristic 0, Coates (see Notes, p. 47) has given an excellent algorithm for constructing $L(D)$, for which there are available some software packages to implement it. It is possible to verify that with some very minor modifications the Coates algorithm also works in the present context of an algebraic curve C defined over a finite field of constants.

Let C, C' be non-singular, irreducible, complete curves and consider a morphism

$$C$$
$$\downarrow \pi$$
$$C'.$$

That is a continuous covering of topological spaces; in particular the pre-image of a closed point $P' \in C'$ is a finite set $\pi^{-1}(P')$ of closed points in

C. We denote by $\pi^*\colon k(C') \to k(C)$ the corresponding embedding of function fields. If $P'_v = (R'_v, m'_v)$ is a closed point of C, we denote by $P_i = (R_i, m_i)$, $1 \le i \le r$, the closed points in C with the corresponding discrete valuations $v_1, \ldots, v_r$. If t is a l.u.p. at P'_v and t_i at P_i, we define as usual e_i to be the ramification index of the valuation v_i as an extension of v, that is, as an element of P_i we have $t = t_i^{e_i}$. With the point P'_v we associate a divisor in C by the prescription

$$\pi^*(P'_v) = \sum_{i=1}^r e_i \cdot P_i.$$

Observe that the degree of this divisor is equal to the degree of the covering:

$$\sum_{i=1}^r e_i \deg P_i = [k(C)\colon k(C')].$$

We will say that the covering π is unramified at P'_v if all $e_i = 1$. We extend by linearity the definition of π^* to the group of all divisors on C':

$$\pi^*\colon \mathrm{Div}(C') \to \mathrm{Div}(C).$$

If f' is a function in $k(C')$ then the divisor $\pi^*(f')$ is simply the principal divisor on C corresponding to the composite function $f' \circ \pi = \pi^*(f)$. The covering π is called separable, when the extension $k(C)/k(C')$ of function fields is separable. When this is the case, a differential $\omega' = f' \, dt'$ in $\Omega_{C'}$ gives rise to a differential

$$\pi^*(\omega') = \pi^*(f') \, d(\pi^*(t'))$$

in Ω_C. Clearly for any $g \in k(C')$ we have $\pi^*(g\omega') = \pi^*(g) \cdot \pi^*(\omega')$. If (ω') denotes the divisor of the differential ω', we want to consider the relation between $\pi^*(\omega')$ and $(\pi^*(\omega'))$; it turns out that the difference between the canonical classes of C and C' is a divisor which describes the ramification behavior of the covering $\pi\colon C \to C'$. Before stating the main result we define the *different*. Let $Q_w = (R_w, m_w)$ be a closed point of C with discrete valuation w which extends the discrete valuation v of the closed point $P_v = (R_v, m_v)$ of C'; let the completions of the function fields be denoted respectively by K_w and K_v; also let $\hat{R}_w$ and $\hat{R}_v$ be the completions of the discrete valuation rings R_w and R_v. Since we are assuming that $k(C)$ is separable over $k(C')$, the trace mapping $tr\colon K_w \to K_v$ is well defined and non-zero. The differential exponent d_w of K_w over K_v is defined as the largest integer d such that $\mathrm{tr}(x) \in R_v$ for all $x \in K_w$ with $w(x) \ge -d$. The main properties of the differential exponent d are ([11], [99]):

(i) $d_w = 0$ for all except a finite number of closed points,
(ii) $d_w = \mathrm{ord}_w(F'(t_w))$ if and only if $R_w = R_v[t_w]$, where

$$F(T) = N_{K_w/K_v}(T - t_w) = T^n + \sum_{i=1}^{n} a_i x^{n-i}.$$

Definition 5.15 *The different of the covering* $\pi: C \to C'$ *is the divisor in* Div(C) *given by*

$$D_{C/C'} = \sum_{Q_w} d_w \cdot Q_w,$$

where Q_w *runs over all the closed points in* C.

When the field of constants k is perfect, as is the case with $k = \mathbb{F}_q$, the divisor $D_{C/C'}$ contains information which is purely of a geometric nature, i.e. its essential nature does not change under a finite separable extension of the field of constants of the function fields $k(C)$ and $k(C')$. In Theorem 5.7 we assume that both curves C and C' are defined over the same field k and their function fields have k as their field of constants.

Theorem 5.7 (Hurwitz–Zeuthem) *Let* $\pi: C \to C'$ *be a separable covering with different* $D_{C/C'}$. *If* ω' *is any differential in* $\Omega_{C'}$, *then we have*

$$(\pi^*(\omega')) - \pi^*((\omega')) = D_{C/C'}.$$

The following is also a useful corollary.

Corollary 5.7.1 *Let* $g(C)$, *resp.* $g(C')$, *be the genus of* C, *resp. of* C'. *Then we have*

$$2g(C) - 2 = (2g(C') - 2)[k(C): k(C')] + \deg D_{C/C'}.$$

Given an abstract curve C with function field $k(C)$, it is possible to find an affine plane model

$$C: f(X, Y) = 0, \qquad f \in k[X, Y],$$

whose only singularities are ordinary double points, i.e. points with distinct tangents. The geometry of such a situation is similar to that of example $C: y^2 = x^3 + x^2$ studied in Sections 4.2. If $P_1, \ldots, P_s$ is the complete set of such double points, then by realizing the function field $k(C) = k(x, y)$ as a separable algebraic extension of degree n of $k(x)$, we can view C as a covering of the projective line $\pi: C \to \mathbb{P}^1$ ramified only at the points $P_1, \ldots,$ P_s. Furthermore, the different is $D_{C/\mathbb{P}^1} = P_1 + \cdots + P_s$; by the corollary above, since the genus of $\mathbb{P}^1$ is 0, we get

$$2g - 2 = n(-2) + s.$$

For example the curve

$$C: y^2 = 4x^3 - g_2 x - g_3, \qquad g_2, g_3 \in k,$$

where $4x^3 - g_2 x - g_3$ has no double root, can be viewed as a covering $\pi: C \to \mathbb{P}^1$ of the projective line of degree $[k(C): k(x)] = 2$ ramified at the point at infinity and at the roots of $4x^3 - g_2 x - g_3$; the degree of the different is in this case 4 and hence from $2g(C) - 2 = 2(-2) + 4$ we obtain that the genus of C is 1. There are many other formulas for computing the genus of $k(C)$ when a suitable affine model for C is given. For example, in any characteristic, if

$$C: f(x, y) = 0, \qquad f \in k[x, y], \quad \deg f = d,$$

is an affine model whose singularities are only ordinary multiple points, i.e. with distinct tangents, say at the point $P_1, \ldots, P_s$ with multiplicities $m_1, \ldots, m_s$, then the genus of C is given by the formula

$$g_C = \frac{(d - 1)(d - 2)}{2} - \sum_{i=1}^{s} \frac{m_i(m_i - 1)}{2}.$$

Recall that a point $P = (0, 0)$ is called an ordinary multiple point of $f(x, y) = 0$ of multiplicity m if the homogeneous part of lowest degree

$$f(x, y) = h_d(x, y) + h_{d+1}(x, y) + \ldots$$

has a factorization $h_d(x, y) = \prod_{i=1}^{d} (\alpha_i x + \beta_i y)$ into distinct linear factors.

Remark Let $k = \mathbb{F}_q$; the curve

$$C: xy^q - yx^q = 1$$

is free of singularities. Since its degree is $q + 1$, its genus is $q(q - 1)/2$. This curve has an interesting history dating back to Dickson who was interested in the invariants of $PGL_2(\mathbb{F}_q)$. Over the algebraic closure $\bar{k}$ of k, the torus $H = \{\lambda \in \bar{k}^\times : \lambda^{q+1} = 1\}$ acts on the curve by homotheties; $SL_2(\mathbb{F}_q)$ acts by a linear change of coordinates. Drinfeld was the first to realize that a study of the unramified coverings of C leads to a construction of some interesting representations of the group $SL_2(\mathbb{F}_q)$, i.e. those which are not induced from representations of parabolic subgroups. The corresponding projective nonsingular curve

$$C: xy^q - x^q y = z^{q+1},$$

is birationally equivalent, over $\mathbb{F}_{q^2}$, to the curve

$$C: x^{q+1} + y^{q+1} + z^{q+1} = 0,$$

on which the unitary group $U_3(\mathbb{F}_q)$ acts. An example of such a curve is the Fermat cubic

$$C\colon x^3 + y^3 + z^3 = 0$$

over the field $\mathbb{F}_4$. These examples and others, which have been considered by Goppa [32], will be used later in Section 5.7.3 to construct interesting codes.

Before closing this section we want to discuss another concept that is useful in the construction of Goppa codes.

5.4.6 Linear series

Let C be an irreducible algebraic curve with function field $k(C)$. Let D be a divisor of degree s and consider the vector space $L(D)$. If V is a vector subspace of $L(D)$, the *linear series* associated to V is the family of all divisors on C given by

$$g\{V\} = \{(f) + D \mid f \in V - \{0\}\}.$$

As is obvious from the definition of $L(D)$, the linear series consists only of positive divisors. In terms of the classical notation, if $\dim V = r + 1$, then $g(V) = g_s^r$. If $V = L(D)$, then $g(V)$ is called a *complete linear series*; it consists of all positive divisors on C linearly equivalent to D. Let $\mathbb{P}^r = \mathbb{P}^r(V)$ denote the projective space associated to V, which is made up of all lines in V through the origin; let $\varphi_0, \varphi_1, \ldots, \varphi_r$ be a basis for V. The mapping

$$g(V) \to \mathbb{P}^r$$

$$(x_0\varphi_0 + \cdots + x_r\varphi_r) + D \to (x_0, \ldots, x_r)$$

sets up a one-to-one correspondence between the linear series and projective r-space. If $V = L(D)$ and $\varphi_0, \ldots, \varphi_r$ is a basis, then the mapping

$$\varphi\colon C \to \mathbb{P}^r, \qquad \varphi(P) = (\varphi_0(P), \ldots, \varphi_r(P))$$

embeds the curve C in projective r-space as a normal curve, i.e. a curve which is not the projection of a curve from a larger projective space, of genus g and degree $s = \deg D$.

The embedding φ is of particular importance in the theory of algebraic curves of genus > 1 which are not hyperelliptic, i.e. double coverings of the projective line ramified at a finite number of points. If C is such a curve and W denotes its canonical class, then, as we have seen above, $L(W) \simeq \Omega_C(\mathcal{O})$, the space of holomorphic differentials on C. If $\omega_1, \ldots, \omega_g$ is a basis, then

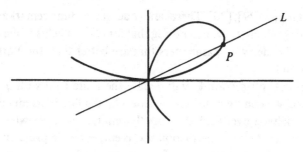

Figure 5.4   The curve $C: x^3 + y^3 - 2xy = 0$.

$$\varphi: C \to \mathbb{P}^{g-1}, \qquad \varphi(P) = (\omega_1(P), \dots, \omega_g(P))$$

provides an embedding of C as a non-singular curve in $\mathbb{P}^{g-1}$.

Example Consider the plane cubic curve $C: x^3 + y^3 - 2xy = 0$ whose only singularity is a double point at the origin with distinct tangents which coincide with the coordinate axis (Figure 5.4). If $L: y = Tx$ is the line passing through the origin and the point P, then the mapping $P \to T$ provides a birational isomorphism between C and $\mathbb{P}^1$; in fact $k(C) \simeq k(T)$ via the mapping $x = 2T/(1 + T^3)$ and $y = 2T^2/(1 + T^3)$. On $\mathbb{P}^1$ let P_0, P_1 and P_∞ correspond respectively to the polynomials $T = 0$, $T - 1 = 0$ and the function $1/T$. If D denotes the divisor

$$D = 2P_0 + P_1 + P_\infty,$$

which is of degree 4, then $L(D)$ is of dimension 5. Clearly any function $f \in k(T)$ which satisfies $(f) + D \geq 0$ is of the form

$$f = \lambda_0 + \lambda_1 T + \lambda_2 \frac{1}{T} + \lambda_3 \frac{1}{T^2} + \frac{\lambda_4}{T-1}, \qquad \lambda_i \in k.$$

In terms of the x, y-coordinates we get as a basis for $L(D)$ on C the five functions

$$\varphi_0 = 1, \qquad \varphi_1 = \frac{y}{x}, \qquad \varphi_2 = \frac{x}{y}, \qquad \varphi_3 = \left(\frac{x}{y}\right)^2, \qquad \varphi_4 = \frac{x}{y-x}.$$

5.5 Algebraic geometric codes

5.5.1 Algebraic Goppa codes

In this section we use the theory of algebraic curves over finite fields as presented in Section 5.4 to define a large class of linear codes first introduced by Goppa [29], [30]. These new codes have many remarkable properties, among which it is worth noting that for certain fields $\mathbb{F}_q$ they provide sequences of codes with parameters that are better than the Varshamov–Gilbert bound.

Throughout this section $\mathbb{F}_q$ will denote the finite field with $q = p^f$ elements. By C we denote a projective, smooth, absolutely irreducible curve of genus g defined over $\mathbb{F}_q$. Let us briefly remind ourselves what all these adjectives mean. As a geometric object C is embedded in projective r-space $\mathbb{P}^r$ as the set of common zeros of a system of s homogeneous forms in $\mathbb{F}_q[x_0, \ldots, x_r]$:

$$f_i(x_0, \ldots, x_r) = 0, \qquad 1 \le i \le s,$$

of dimension 1. For instance a plane projective curve is defined by a single homogeneous form

$$f(x_0, x_1, x_2) = 0.$$

Smooth means that at each point of C there is a uniquely determined tangent direction. This is a condition which for a plane algebraic curve can be checked easily as follows: let $P = (x_0, y_0, z_0)$ be a point on C and let

$$f(x, y, z) = f_1(x, y, z) + f_2(x, y, z) + \cdots + f_d(x, y, z)$$

be the decomposition of the Taylor expansion of f about the point $P = (x_0, y_0, z_0)$ into homogeneous terms f_i of degree i. The condition of smoothness at P is then that the linear term

$$f_1(x, y, z) = f_x(P)(x - x_0) + f_y(P)(x - y_0) + f_z(P)(z - z_0),$$

should not be identically zero. Here $f_z = \partial f/\partial z$ etc. The condition of absolute irreducibility guarantees that the curve is connected in a certain sense. In fact, if C is defined as above by a homogeneous form f in $\mathbb{F}_q[x, y, z]$, irreducibility simply means that f is not the product of two non-constant homogeneous forms in $\mathbb{F}_q[x, y, z]$ of lower degree. Absolute irreducibility is a geometric property, which means that f is irreducible over any finite extension of $\mathbb{F}_q$, i.e. the curve C when viewed over the algebraic closure of $\mathbb{F}_q$ is not the disjoint union of two other curves. In practical terms, when C is defined by an affine model $f(x, y)$, absolute irreducibility implies that the coordinate ring $\mathbb{F}_q[x, y]/(f)$ is an integral domain and remains so when the field $\mathbb{F}_q$ is replaced by any finite extension. This in turn guarantees that the field of quotients is a function field of transcendence degree 1. As for the genus of C we remind the reader that it is a measure of the complexity of the curve C as compared with the projective line. For example if C is a plane projective curve defined by a homogeneous form $f(x, y, z) = 0$ of degree d, then from the point of view of birational geometry C looks like the projective line $\mathbb{P}^1$, i.e. they both have isomorphic function fields, if the only singularities of C are $(d - 1)(d - 2)/2$ double points. The genus of C can be considered as the deficiency in the number of double points from the total allowable number.

Let $P_1, \ldots, P_n$ be n distinct points on C and consider the divisor

$$D = P_1 + \cdots + P_n.$$

Let G be also a positive divisor on C whose support is not contained in the set $\{P_1, \ldots, P_n\}$, that is $G = \sum_Q m_Q \cdot Q$, $m_Q \geq 0$ and the points that appear in G with $m_Q > 0$ are distinct from the P_i. Let

$$\Omega_C(D - G) = \{\omega \in \Omega_C : (\omega) + D - G \geq 0\};$$

this is the space of differentials which have zeros at the points Q in the support of G with multiplicities at least m_Q and which are regular outside the set $\{P_1, \ldots, P_n\}$ where they are allowed to have at most simple poles. We have already noted in Section 5.4 that this space is isomorphic over $\mathbb{F}_q$ to the space $L(K + D - G)$. Hence its dimension over $\mathbb{F}_q$, as calculated by the Riemann–Roch theorem, is

$$\dim \Omega_C(D - G) = \dim L(K + D - G)$$

$$= \deg(K + D - G) + 1 - g + \dim L(G - D).$$

If we introduce the assumption

(A) $$\deg(G - D) < 0,$$

in which case the space $L(G - D)$ is empty, then the dimension becomes

$$g - 1 + \deg D - \deg G.$$

To a differential ω in $\Omega_C(D - G)$ we associate the residue mapping

$$(\mathrm{res}_{P_1} \omega, \ldots, \mathrm{res}_{P_n} \omega),$$

where $\mathrm{res}_{P_i} \omega$ is the residue of ω at the point P_i as defined in Section 5.4. The residue mapping gives rise to a sequence of linear mappings

$$0 \to \Omega_C(-G) \to \Omega_C(D - G) \xrightarrow{\varphi} \mathbb{F}_q^n,$$

where $\Omega_C(-G)$ is the kernel of φ, i.e. the subspace of $\Omega_C(D - G)$ consisting of differentials which have no poles at the points $P_1, \ldots, P_n$; since $\Omega_C(-G) \simeq L(K - G)$, the kernel of the residue mapping is empty if the following extra assumption is made:

(B) $$\deg(K - G) < 0,$$

that is to say $2g - 2 < \deg G$. The image of the residue mapping then defines a subspace of $\mathbb{F}_q^n$ isomorphic to $\Omega_C(D - G)$.

Suppose the differential ω has exactly d poles, say at the points $P_1, \ldots, P_d$, i.e. the vector

$$\varphi(\omega) = (\mathrm{res}_{P_1}\omega, \ldots, \mathrm{res}_{P_n}\omega)$$

has d non-zero components. Then clearly ω belongs to the subspace

$$\Omega_C(P_1 + \cdots + P_d - G)$$

which is isomorphic to $L(K + P_1 + \cdots + P_d - G)$; this latter space has then to be non-empty and we must have

$$\deg(K + P_1 + \cdots + P_d - G) = \deg K + \sum_{i=1}^{d} \deg P_i - \deg G \geq 0;$$

hence the following inequality for the minimum number of poles of a differential ω in $\Omega_C(D - G)$ must hold

$$d \cdot \max_{1 \leq i \leq n} (\deg P_i) \geq \deg G - 2g + 2.$$

Definition 5.16 *Let C be a smooth, projective, absolutely irreducible algebraic curve defined over $\mathbb{F}_q$ of genus g. Let $D = P_1 + \cdots + P_n$, G be two positive divisors on C with disjoint supports and with $P_1, \ldots, P_n$ distinct. Suppose that*

$$2g - 2 < \deg G < \deg D.$$

Then the image of the residue mapping

$$\Omega_C(D - G) \to \mathbb{F}_q^n$$

$$\omega \to (\mathrm{res}_{P_1}\omega, \ldots, \mathrm{res}_{P_n}\omega)$$

is an algebraic Goppa code, denoted by $\Gamma_C(D, G)$. The dimension of $\Gamma_C(D, G)$ is $g - 1 + \deg D - \deg G$ and its relative distance d satisfies $d \cdot \max_{1 \leq i \leq n} \times (\deg P_i) \geq \deg G - 2g + 2$.

A second approach to the definition of alagebraic Goppa codes is provided by the Residue theorem, which gives a dual description via the parity check matrix. To simplify the notation we make the following assumption:

(C)   All the points $P_1, \ldots, P_n$ are of degree 1 over $\mathbb{F}_q$.

Definition 5.17 *With the notations and assumptions as above, let $\varphi_1, \ldots, \varphi_s$ be a basis over $\mathbb{F}_q$ of $L(G)$. Then we have that the matrix*

$$H = (\varphi_i(P_j)), \qquad 1 \le i \le s$$

$$1 \le j \le n$$

is the parity check matrix of the code $\Gamma_C(D, G)$.

To establish the equivalence of the above two definitions we prove the following lemma, which is a simple corollary of the Residue theorem.

Lemma 5.3 *Let* $\varphi \in L(G)$ *and* $\omega \in \Omega_C(D - G)$. *Then we have*

$$\sum_{i=1}^{n} \varphi(P_i) \operatorname{res}_{P_i}(\omega) = 0.$$

Proof. Since $(\varphi) + G \ge 0$ and $(\omega) + D - G \ge 0$, we have $(\varphi\omega) + D \ge 0$ and hence the poles of $\varphi\omega$ are among the points $P_1, \ldots, P_n$ and they are at most simple. As each of the points $P_1, \ldots, P_n$ are of degree 1, i.e. each $\varphi(P_i)$ lies in $\mathbb{F}_q$, we have $\operatorname{res}_{P_i}(\varphi\omega) = \varphi(P_i) \operatorname{res}_{P_i}(\omega)$. Therefore the Residue theorem implies

$$0 = \sum_{P} \operatorname{res}_P(\varphi\omega)$$

$$= \sum_{i=1}^{n} \operatorname{res}_{P_i}(\varphi\omega)$$

$$= \sum_{i=1}^{n} \varphi(P_i) \operatorname{res}_{P_i}(\omega).$$

This proves the lemma.

The equivalence of the two definitions is now clear since a code word $c = (\operatorname{res}_{P_1}\omega, \ldots, \operatorname{res}_{P_n}\omega)$ in $\Gamma_C(D, G)$ is, according to the lemma, in the null space of the parity check matrix $H = (\varphi_i(P_j))$.

Remark It is possible to modify the definition above so as to include the case where not all the points $P_1, \ldots, P_n$ are of degree 1. For example we could assume that the residue fields of the P_i are all contained in $\mathbb{F}_{q^m}$. A code could then be defined by considering the image of the residue mapping $\Omega(D - G) \to (\mathbb{F}_{q^m})^n$ as a vector space over $\mathbb{F}_q$. The dual definition would correspond to the parity check matrix $H = (\varphi_i(P_j))$ where the φ_i are a basis of $L(G)$ over $\mathbb{F}_{q^m}$. In this setting the basic parameters can still be calculated from the Riemann–Roch theorem. This variation, which was also introduced by Goppa ([30], p. 82, §5) under the name of (D, G, m)-code, may prove to be useful when dealing with the problem of constructing sequences of binary codes with parameters which go beyond the Varshamov–Gilbert bound.

Remark The dimension of $\Gamma_C(D, G)$ can be calculated from the second definition by observing that since $2g - 2 < \deg G$, $\dim_{\mathbb{F}_q} L(G) = \deg G + 1 - g$ is the rank of H. Hence the dimension of the solution space is $n + g - 1 - \deg G$.

5.5.2 Codes with better rates than the Varshamov-Gilbert bound. Calculation of Parameters

Let k (resp. d) be the dimension (resp. distance) of $\Gamma_C(D, G)$. The basic parameters of the algebraic Goppa code are the *transmission rate* $R = k/n$ and the *relative distance* $\delta = d/n$. For the code $\Gamma_C(D, G)$ defined in Section 5.1 these rates satisfy

$$R = \frac{n - \deg G + g - 1}{n}$$

$$\delta \geq \frac{\deg g - 2g + 2}{n}.$$

If we choose $P_1, \ldots, P_n, Q$ to be all the points on the curve C rational over $\mathbb{F}_q$, i.e. the points on C of degree 1, and if α denotes an integer satisfying

$$2g - 2 < \alpha < n,$$

then the parameters of the corresponding algebraic Goppa code $\Gamma_C(D, G)$, $(D = P_1 + \cdots + P_n, G = \alpha Q)$, satisfy

$$R + \delta \geq 1 - \gamma_C,$$

where

$$\gamma_C = (g_C - 1)/(N_C - 1),$$

g_C is the genus of C and N_C is the number of points on C rational over $\mathbb{F}_q$.

Weil's proof of the Riemann hypothesis for the function field of C provides the upper bound

$$N_C \leq q + 1 + 2g_C q^{1/2} \tag{5.8}$$

from which it follows that

$$\frac{1}{2q^{1/2}} \leq \frac{q^{1/2} + 2}{2(N_C - 1)} + \gamma_C. \tag{5.9}$$

The realization that there is a sequence of curves C for which it is possible to improve asymptotically the lower bound (5.9) has led to the proof of the following result.

Theorem 5.8 *Let $q = p^{2f} \geq 49$. For each prime $l > p$ let $C_l = X_0(l)$ be the modular curve which parametrizes elliptic curves with a subgroup of order l. Let g_l be the genus of C_l and N_l the number of points on C_l rational over $\mathbb{F}_q$. Then we have for $\gamma(C_l) = (g_l - 1)/(N_l - 1)$*

$$\lim_{l \to \infty} \gamma(C_l) = \frac{1}{\sqrt{q} - 1}$$

Furthermore, if $\varphi(\delta) = \delta \log_q(q - 1) - \delta \log_q \delta - (1 - \delta) \log_q(1 - \delta)$ denotes the entropy function, then the equation

$$\varphi(\delta) - \delta = \frac{1}{\sqrt{q} - 1}$$

has two distinct roots δ_1, δ_2 and hence it is possible to construct algebraic Goppa codes with transmission rate and relative distance which are better than the Varshamov–Gilbert bound in the interval δ_1, δ_2 (see Figure 5.5).

Remark This theorem is due to Tsfasman, Vladut and Zink [90] in the case $q = p^2$ or $q = p^4$. The general case $q = p^{2f}$ was also proved by the same authors (see Manin & Vladut [55] using the theory of Drinfeld modules and independently by Ihara [40]. The result represents a major breakthrough after twenty years of work on the problem of constructing codes with rates which are better than the Varshamov–Gilbert bound.

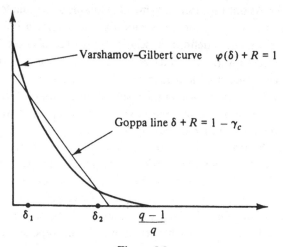

Varshamov–Gilbert curve $\varphi(\delta) + R = 1$

Goppa line $\delta + R = 1 - \gamma_c$

$\delta_1 \qquad \delta_2 \qquad \dfrac{q-1}{q}$

Figure 5.5

Below we shall give several examples and in the rest of the chapter we shall give a brief introduction to the theory of modular curves leading up to a new proof of the theorem based on the Eichler–Shimura congruence relation and the Eichler–Selberg trace for the Hecke operators acting on the space of differentials of the first kind on the curve $X_0(l)$.

5.6 The theorem of Tsfasman, Vladut and Zink

5.6.1 Modular Curves

As already indicated in Section 5.5, the construction of Goppa codes on modular curves lead to some very remarkable codes whose rates are better than those prescribed by the Gilbert–Varshamov bound. The aim of the present section is to give an introduction to the theory of modular curves in so far as they apply to the proof of the theorem of Tsfasman, Vladut and Zink (Section 5.6.9).

   Modular curves are one of the central themes of modern number theory; their usefulness in the study of modular forms, abelian fields, diophantine equations etc. is well documented in the literature. For reasons of space our treatment will be rather brief. In fact we shall try to emphasize those aspects of the theory which are of a classical nature and that are easy to comprehend with a minimum knowledge of algebraic geometry. Our primary goal has been to make available to the coding theorists an outline of the main applications of modular curves to the construction of Goppa codes; we do hope that the interested reader will find here enough material to facilitate his approach to the rather extensive and complicated literature on the subject. As our program is to make available a formula for counting the number of rational points on a modular curve over a finite field, we shall first develop some definitions leading up to Igusa's description of a model for the curve $X_0(N)$ defined over the integers $\mathbb{Z}$ with good reduction modulo every prime p which does not divide the level N; this will be followed by a brief description of the space of differentials of the first kind on $X_0(N)$ as modular forms of weight two on the group $\Gamma_0(N)$ and will then recall the basic result of Eichler and Shimura relating the trace of the Hecke operators to the trace of Frobenius acting on the l-primary part of the torsion points of the Jacobian variety of $X_0(N)$; finally we will give the formula of Eichler and Selberg for the trace of the Hecke operators, thus completing the new proof of the theorem of Tsfasman, Vladut and Zink. The original proof is based on a lower bound for the number of points on $X_0(N)$, rational over $\mathbb{F}_{p^2}$, which correspond to the supersingular elliptic curves.

5.6.2 Elliptic curves over $\mathbb{C}$

Let H be the Poincaré upper half plane, i.e. $\{z = x + iy \in C : y > 0\}$. The modular group Γ consisting of two by two integer matrices $\sigma = \begin{pmatrix} a & b \\ c & d \end{pmatrix}$ with $\det(\sigma) = 1$ acts directly on H via linear fractional transformations

$$\sigma(z) = \frac{az + b}{cz + d}.$$

In the following we shall identify Γ with the quotient $SL_2(\mathbb{Z})/\{\pm 1_2\}$. Under the above action the orbits of the group Γ can be identified with points in the fundamental domain

$$D = \{z = x + iy \in H : |x| \leq \tfrac{1}{2}, |z| \geq 1\}.$$

For this identification to be precise, only half of the boundary of D should be included so as to account for the action of the two generators of Γ:

$$T(z) = z + 1, \qquad S(z) = -\frac{1}{z}.$$

By abuse of notation we write $D = H/\Gamma$. D can be made into a compact topological space by adding the 'point at infinity' ∞ and prescribing that any subset D_Y of D of the form $\{z = x + iy : |x| \leq \tfrac{1}{2}, y \geq Y\}$ shall be an open neighborhood of ∞. The resulting space is a compact Riemann surface of genus 0, i.e. an algebraic curve over $\mathbb{C}$ whose field of meromorphic functions is isomorphic to that of the projective line $\mathbb{P}^1(\mathbb{C})$. As we shall see below, this correspondence gives the equality

$$\mathbb{P}^1(\mathbb{C}) = H/\Gamma \cup \{\infty\}$$

and also permits us to think of $\mathbb{P}^1(\mathbb{C})$ as the parameter space for the family of *isomorphism classes of elliptic curves* over $\mathbb{C}$, with the understanding that the point ∞ corresponds to the singular cubic curve $y^2 = 4x^3$. Let us try to clarify the situation. There are many ways of defining an an elliptic curve; for the purpose at hand it may be best to think of an elliptic curve as a one-dimensional torus, i.e. the quotient of $\mathbb{C}$ by a lattice $L = \mathbb{Z} + \mathbb{Z}z$:

$$E(z) = \mathbb{C}/L,$$

where z is a complex number with positive imaginary part. The classical and well-known theory of the doubly periodic Weierstrass p-function

$$p(z; L) = \frac{1}{z^2} + \sum_{\omega} \left\{ \frac{1}{(z - \omega)^2} - \frac{1}{\omega^2} \right\}, \qquad \omega \in L - \{0\},$$

provides an analytic isomorphism

$$\mathbb{C}/L \xrightarrow{\sim} E$$

between $E(z)$ and the Weierstrass model

$$E: y^2 = 4x^3 - g_2 x - g_3;$$

recall that the map which sends a point $u \in \mathbb{C}/L$ to a point on E with coordinates (x, y) is that given by $u \to (p(u), p'(u))$. The addition formulas for p and p' are simply a reflection of the fact that E inherits an abelian group structure from $\mathbb{C}/L$ via this isomorphism. The coefficients g_2 and g_3 depend on z and are such that the discriminant

$$\Delta = g_2^3 - 27 g_3^2$$

remains bounded and non-zero in D. The function $q = \exp(2\pi i z)$, which maps an open neighborhood D_Y of the point ∞ onto a disc, serves as a local uniformizing parameter. To the elliptic curve $E(z)$, $z \in H$, there is associated a function $j(z)$ which is an invariant of the isomorphism class of $E(z)$. In terms of the local uniforming parameter q, the expansion of j is given by

$$j(z) = \left\{ 1 + 240 \sum_{n=1}^{\infty} \sigma_3(n) q^n \right\}^3 \Big/ q \prod_{n=1}^{\infty} (1 - q^n)^{24},$$

where $\sigma_3(n)$ is the sum of the cubes of the positive divisors of n. Many of the coefficients of this remarkable function have been calculated; the formal q-expansion of j when considered in the ring $\mathbb{F}_q((q))$ permits the calculation of an equation for the curve $X_0(N)$ for any N not divisible by p. When the elliptic curve E is given by an equation of the form $y^2 = 4x^3 - g_2 x - g_3$, the j-invariant is defined algebraically by the formula

$$j(E) = (12)^3 g_2^3 / (g_2^3 - 27 g_3^2).$$

An important property of the j-invariant is the fact that two elliptic curves E and E' with $j(E) = j(E')$ are isomorphic; the isomorphism need not be defined over the field which contains the coefficients g_2, g_3. Each point $z \in D$ then defines an elliptic curve $E(z)$ with finite j-invariant $j(z)$. To make the correspondence complete we observe that if a value of $j = j(z)$ is given then it is possible to write an equation for a representative curve in the isomorphism class of elliptic curves with the given j-invariant; namely,

(i) If $j \neq 0, (12)^3$, then $E: y^2 = x^3 - \dfrac{27j}{j - 1728}(x + 1)$

(ii) If $j = 0$, then $E: y^2 = 4x^3 - 1$

(iii) If $j = 1728$, then $E: y^2 = 4x^3 - 3x$.

As the point $z \in H$ approaches the point ∞, the value of $j(z)$ become unbounded, thus reflecting the fact that $E(z)$ degenerates into a plane cubic curve with a singularity for which a model can be chosen to be $y^2 = 4x^3$.

To recapitulate what we have just said, we observe that two points z and z' in H related by a unimodular transformation $\sigma \in \Gamma$: $z' = \sigma(z)$ give rise to two elliptic curves $E(z)$ and $E(z')$ which are birationally isomorphic, i.e. they have isomorphic function fields. On the other hand the function $j(z)$ defined on H is invariant under Γ: $j(z') = j(z)$ and its value characterizes the isomorphism class of the elliptic curve $E(z)$. The projective line $\mathbb{P}^1(\mathbb{C})$, with the field of rational functions $\mathbb{C}(j)$ as its function field, can now be considered as a parameter space for the isomorphism class of elliptic curves over $\mathbb{C}$; in this sense we shall refer to $\mathbb{P}^1(\mathbb{C})$, or equivalently $X_0(1) = H/\Gamma \cup \{\infty\}$, simply as the *j-line*.

Remark Over the complex numbers, the classical theory of elliptic functions and elliptic curves, the rudiments of which were introduced above, is very rich in special functions and relations. This makes possible a study by means of explicit formulas. In this situation, and perhaps more interestingly, the classical theory suggests the consideration of the q-expansions over fields other than $\mathbb{C}$ and specially over fields of characteristic not 0. To carry out this suggestion rigorously, Igusa [38] had to rework the theory algebraically as well as geometrically; Igusa's main theorem (see Section 5.6) provides a sort of universal model for the parameter spaces of the families of elliptic curves with a so-called level N-structure. The difficulty in the analysis is due in part to problems having to do with char 2 and 3 and the fact that certain elliptic curves have more automorphisms than they should have.

We now continue with the description of the moduli spaces corresponding to congruence subgroups of Γ; these will correspond to finite algebraic extensions of the function field $\mathbb{C}(j)$ of the j-line which possess many pleasant properties.

Let N be a positive integer and let $E = \mathbb{C}/L$ be an elliptic curve with period lattice $L = \mathbb{Z}\omega_1 + \mathbb{Z}\omega_2$, where the generators are required to satisfy $\omega_2/\omega_1 \in H$. Let C_N be a cyclic subgroup of E of order N, which without loss of generality we may identify with the integral multiples of ω_2/N modulo L. Under the action of the unimodular transformation $\sigma = \begin{pmatrix} a & b \\ c & d \end{pmatrix}$ in Γ, the lattice L goes into the lattice $L = \mathbb{Z}\omega'_1 + \mathbb{Z}\omega'_2$, where $\omega'_1 = a\omega_1 + b\omega_2$, $\omega'_2 = c\omega_1 + d\omega_2$; the generator of C_N goes over into

$$\frac{\omega'_2}{N} \equiv \frac{c\omega_1 + d\omega_2}{N} \bmod L.$$

From this relation, it is clear that under σ the group C_N remains invariant if and only if $c \equiv 0 \bmod N$. The elements σ which satisfy this requirement make up the subgroup

$$\Gamma_0(N) = \left\{ \sigma = \begin{pmatrix} a & b \\ c & d \end{pmatrix} \in \Gamma : c \equiv 0 \bmod N \right\}.$$

The index of this group in $\Gamma (= \Gamma_0(1))$ is given by the formula

$$[\Gamma : \Gamma_0(N)] = N \prod_{p|N} \left(1 + \frac{1}{p} \right).$$

The quotient $Y_0(N) = H/\Gamma_0(N)$, i.e. the space of orbits in H under the action of $\Gamma_0(N)$, can be thought of as a covering of H/Γ which parametrizes isomorphism classes of pairs (E, C) consisting of an elliptic curve E together with a fixed cyclic subgroup C of E of order N. The space $Y_0(N)$ is the complement of a finite number of points of a compact Riemann surface; to compactify $Y_0(N)$, one then adds the missing points to $H/\Gamma_0(N)$ with appropriate open neighborhoods. The missing points are the cusps of $\Gamma_0(N)$; these correspond to equivalence classes of points on the projective line $\mathbb{P}^1(\mathbb{Q}) = \mathbb{Q} \cup \{\infty\}$ under the action of $\Gamma_0(N)$. A useful set of representatives for the orbits $\mathbb{P}^1(\mathbb{Q})/\Gamma_0(N)$ is described as follows. Consider the set of all tuples c, d of positive integers satisfying

$$(c, d) = 1, \quad d|N, \quad 0 < c \le N/d$$

(or c in any set of representatives for $\mathbb{Z} \bmod (N/d)$).

If Γ_s is the stabilizer of any cusp s in $\mathbb{P}^1(\mathbb{Q})/\Gamma_0(N)$, then the cardinality of $\mathbb{P}^1(\mathbb{Q})/\Gamma_0(N)$ is the same as the cardinality of the double cosets $\Gamma_0(N) \backslash \Gamma / \Gamma_s$. If we take for s the cusp 0, then this cardinality is the number of pairs $\{c, d\}$ as above modulo the equivalence defined by $\{c, d\} \sim \{c', d'\}$ if there are matrices $\sigma' = \begin{pmatrix} a' & b' \\ c' & d' \end{pmatrix}$ and $\sigma = \begin{pmatrix} a & b \\ c & d \end{pmatrix}$ in Γ satisfying $\sigma' = \sigma \begin{pmatrix} 1 & 0 \\ m & 1 \end{pmatrix}$ for some $m \in \mathbb{Z}$. Thus for a fixed d the number of inequivalent cusps is $\varphi((d, N/d))$ (φ = Euler's function), and hence the number of cusps for $\Gamma_0(N)$ is

$$v_\infty = \sum_{d|N} \varphi((d, N/d)).$$

In particular if N is a prime number l then $v_\infty = 2$. The compactified Riemann surface

$$X_0(N)_C = H/\Gamma_0(N) \cup \mathbb{P}^1(\mathbb{Q})/\Gamma_0(N),$$

corresponds to the set of complex points on a typical *modular curve*: $X_0(N)$.

The interpretation of $X_0(N)$ as a moduli (or parameter) space now requires that the cusps (i.e. the points at infinity) be associated with 'generalized elliptic curves', i.e. curves which are no longer of genus 1 and where a group law can be defined and a distinguished cyclic group of order N can be isolated. For the purpose at hand the most important property of $X_0(N)$ is the fact, first proved by Igusa [38], that $X_0(N)$ has a non-singular projective model which is defined by equations over $\mathbb{Q}$ whose reduction modulo primes p, $p \nmid N$, are also non-singular. Before we give an example of another interesting modular curve we remark that a well-known calculation based on the Riemann–Hurwitz formula for the covering $H/\Gamma_0(N) \to H/\Gamma$ gives that the genus of $X_0(N)$ is (Shimura [79], p. 25)

$$g_0(N) = 1 + \frac{\mu}{12} - \frac{v_2}{4} - \frac{v_3}{3} - \frac{v_\infty}{2},$$

where $\mu = [\Gamma : \Gamma_0(N)]$, v_∞ is the number of cusps and

$$v_2 = \begin{cases} 0 & \text{if } N \text{ is divisible by } 4 \\ \prod_{p \mid N}\left(1 + \left(\frac{-1}{p}\right)\right) & \text{otherwise} \end{cases}$$

$$v_3 = \begin{cases} 0 & \text{if } N \text{ is divisible by } 9 \\ \prod_{p \mid N}\left(1 + \left(\frac{-3}{p}\right)\right) & \text{otherwise.} \end{cases}$$

Here $(-1/p) = 0, 1, -1$ (resp. $(-3/p) = 0, 1, -1$) depending on whether $p = 2$, $p \equiv 1 \bmod 4$ or $p \equiv 3 \bmod 4$ (resp. $p = 3$, $p \equiv 1 \bmod 3$ or $p \equiv 2 \bmod 3$). Observe that if $N = l$ is a prime with $-1, -3 \notin \mathbb{F}_l^2$ then $g_0(l) = \frac{1}{12}(l + 1)$.

Again we let N be a positive integer and consider an elliptic curve $E = \mathbb{C}/L$ whose period lattice $L = \mathbb{Z}\omega_1 + \mathbb{Z}\omega_2$ is selected so that $\omega_2/\omega_1 \in H$. Let P be a point on E of order N, which without loss of generality we may take to be $P = \omega_2/N$. Under the action of an element $\sigma = \begin{pmatrix} a & b \\ c & d \end{pmatrix}$ in Γ, the generators of the lattice are transformed into $\omega_1' = a\omega_1 + b\omega_2$, $\omega_2' = c\omega_1 + d\omega_2$. The point P goes over to the point

$$\frac{\omega_2'}{N} \equiv \frac{c\omega_1 + d\omega_2}{N} \bmod L.$$

Now the point P is left fixed by σ if and only if $c \equiv 0 \bmod N$ and $d \equiv 1 \bmod N$; this is precisely the condition requiring that $\begin{pmatrix} a & b \\ c & d \end{pmatrix}$ be an

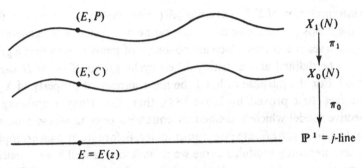

Figure 5.6

element of the congruence subgroup

$$\Gamma_1(N) = \left\{ \begin{pmatrix} a & b \\ c & d \end{pmatrix} \in \Gamma : a \equiv d \equiv \pm 1 \bmod N, c \equiv 0 \bmod N \right\}.$$

The orbits of points in H under $\Gamma_1(N)$ correspond to isomorphism classes of pairs (E, P) consisting of an elliptic curve E together with a point P on E of order N. The quotient $Y_1(N) = H/\Gamma_1(N)$ can also be compactified by adding the cusps, i.e. the orbits of $\mathbb{P}^1(\mathbb{Q})$ under $\Gamma_1(N)$, and thus obtain a complete Riemann surface

$$X_1(N)_\mathbb{C} = H/\Gamma_1(N) \cup \mathbb{P}^1(\mathbb{Q})/\Gamma_1(N)$$

representing the complex points on an algebraic curve which possesses a non-singular projective model defined over $\mathbb{Q}$, with good reduction modulo primes p, $p \nmid N$. Its genus is also given by a formula similar to that for $g_0(N)$. The moduli interpretation of the complete curve $X_1(N)$ requires the consideration of generalized elliptic curves to account for the points at infinity. In a certain sense $X_1(N)$ is a more pleasant object to study; for example it is a Galois covering of the projective line $\mathbb{P}^1$ and hence also of $X_0(N)$.

We can summarize our remarks about the j-line, $X_0(N)$ and $X_1(N)$ by the Figure 5.6. In it the map π_1 projects the point $(E, P) \in X_1(N)$ consisting of the elliptic curve E and the point P of order N onto the point $(E, C) \in X_0(N)$ consisting of E and the cyclic subgroup C generated by P.

5.6.3 Elliptic curves over the fields $\mathbb{F}_p$, $\mathbb{Q}$

To discuss the function fields of the curves $X_1(N)$ and $X_0(N)$ over an arbitrary field k it is useful to consider models for elliptic curves which avoid some of the peculiarities due to characteristic 2 or 3. In the following an elliptic curve defined over a field k will be an irreducible, non-singular, projective algebraic curve of genus 1 together with a distinguished point $\mathcal{O}$, rational over k, which plays the role of the neutral element for the group law. An affine model for such an elliptic curve defined over a field k is given by

$$E: y^2 + a_1 xy + a_3 y = x^3 + a_2 x^2 + a_4 x + a_6, \qquad a_i \in k,$$

with the proviso that the point at infinity $\mathcal{O} = (\infty, \infty)$ is the origin for the group law. Such an equation for E is called a Weierstrass model because in characteristic $p > 3$, the affine coordinate change

$$x \to p = x + \frac{a_1^2 + 4a_2}{12}$$

$$y \to p' = 2y + a_1 x + a_3$$

transforms it into the more familiar equation

$$(p')^2 = 4p^3 - g_2 p - g_3.$$

When a Weierstrass model is given for the elliptic curve E, its discriminant is

$$\Delta(E) = -b_2^2 b_8 - 8b_4^3 - 27b_6^2 + 9b_2 b_6$$

and the j-invariant is

$$j(E) = c_4^3 / \Delta(E),$$

where

$$b_2 = a_1^2 + 4a_2, \qquad b_6 = a_3^2 + 4a_6,$$

$$b_4 = a_1 a_3 + 2a_4, \qquad b_8 = b_2 a_6 - a_1 a_3 a_4 + a_2 a_3^2 - a_4^2,$$

$$c_4 = b_2^2 - 24b_4.$$

Over $\mathbb{C}$ an invariant differential form of the first kind for the elliptic curve $E = \mathbb{C}/L, L = \mathbb{Z} + \mathbb{Z}z$, is given by $dz = dp(z)/p'(z)$; the equivalent object for a Weierstrass model of E is given by

$$\omega = \frac{dx}{2y + a_1 x + a_3}$$

If the Weierstrass model for E is transformed to

$$E': y'^2 + a_1' x'y' + a_3' y' = x'^3 + a_2' x'^2 + a_4' x' + a_6'$$

under the affine coordinate change

$$x \rightarrow u^2 x' + r,$$

$$y \rightarrow u^3 y' + u^2 x' + t,$$

where $u, r, t \in k$ and $u \in k^\times$, then its invariants are also transformed in the following way:

$$j(E) = j(E'),$$

$$\Delta(E) = u'^{12} \Delta(E'),$$

$$\omega \rightarrow u\omega'.$$

As remarked earlier, the invariant $j(E)$ serves as a parameter for the isomorphism class of E; in fact if k is algebraically closed, then a necessary and sufficient condition for two elliptic curves E and E' to be isomorphic is that $j(E) = j(E')$; when k is not algebraically closed, a quadratic extension of k often suffices to define the isomorphism.

The group law on a Weierstrass model E is given explicitly by the Euler method whereby '3 colinear points add up to the zero point $\mathcal{O}$'. In fact, taking $\mathcal{O} = (\infty, \infty)$ as the origin for the group law we have

$$P_1 + P_2 = \mathcal{O},$$

where $P_1 = (x_1, y_1)$, $P_2 = (x_2, y_2)$ if and only if $x_1 = x_2$ and $y_1 + y_2 + a_1 x_1 + a_3 = 0$; otherwise

$$P_1 + P_2 = P_3,$$

with the coordinates of $P_3 = (x_3, y_3)$ given by

$$x_3 + x_1 + x_2 = \lambda^2 + a_1 \lambda - a_2$$

$$-y_3 - a_3 - a_1 x_1 = \lambda x_3 + v$$

and $y = \lambda x + v$ is the equation of the line joining the points P_1 and P_2.

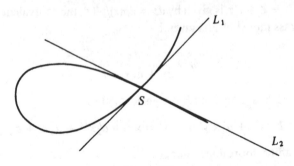

Figure 5.7

When E is a singular plane cubic curve, i.e. $\Delta(E) = 0$, say with a node or a cusp, it is still possible to define a group law on the non-singular locus $E_{ns} = E - \{\text{singular point}\}$ by the following procedure:

(i) $\Delta = 0, c_4 \neq 0$, hence $j(E) = \infty$, then the singular point S is a node and is rational over k. Let $L_i: y = \alpha_i x + \beta_i, i = 1, 2$, be the two distinct tangents passing through S (see Figure 5.7). In this case there is an isomorphism between the non-singular locus $E_{ns} = E - \{S\}$ and the multiplicative group G_m:

$$E_{ns} \xrightarrow{\sim} G_m$$

given by $P = (x, y) \rightarrow (y - \alpha_1 x - \beta_1)/(y - \alpha_2 x - \beta_2)$. If the $\alpha_i \in k$ then clearly this isomorphism is defined over k and for the points with k-valued coordinates we have $E_{ns}(k) \simeq k^{\times}$. Otherwise, as is easily verified, the α_i are defined over a quadratic extension of k and for the k-valued points we have $E_{ns}(k) \simeq \{x \in k(\alpha_1): N_{R(\alpha_1)/k}(x) = 1\}$.

(ii) $\Delta = 0, c_4 = 0$, hence $j(E)$ is indeterminate. Then the singular point S is a cusp. If S is rational over k, as is the case if k is perfect or char$(k) \neq 2$, 3, then one has an isomorphism between the non-singular locus $E_{ns} = E - \{S\}$ and the additive group G_a:

$$E_{ns} \simeq G_a$$

given by $P \rightarrow 1/(m_{P_s} - m_S)$, where m_{P_s} is the slope of the line through P and S and m_S is the slope of the tangent at S (see Figure 5.8). In this case we have for the k-valued points $E_{ns}(k) \simeq k^+$.

Example E: $y^2 - y = x^3 - x^2$. With the aid of the above formulas it is possible to verify that the points on E rational over $\mathbb{F}_{29}$ form an abelian group $E(\mathbb{F}_{29})$ of order $29 + 1$ whose primary decomposition is $(2, 3, 5)$. $E(\mathbb{F}_{31})$ is cyclic of order 25. In Section 5.5 we saw that over $\mathbb{F}_{11}$, E has a node singularity and $E_{ns}(\mathbb{F}_{11}) = \mathbb{F}_{11}^{\times}$.

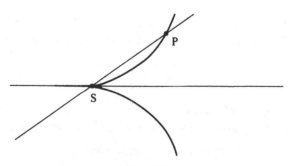

Figure 5.8

Remark An important feature of the above explicit formulas is their universality in the sense that they are defined by equations with integer coefficients. As is now the custom in algebraic geometry this permits us to replace the field of definition k by a more general base (of coefficients), say $\mathbb{Z}$, and still have some control over the possible denominators that may arise in the definition of the various invariants as well as in the group law.

5.6.4 Torsion points on elliptic curves

As is to be expected, the points of finite order on an elliptic curve, and particularly those of order N, play a decisive role in the study of the modular curves $X_0(N)$ and $X_1(N)$. Below we have collected a few basic facts about the group of m-division points on E. Recall that this group is defined as the kernel of the homomorphism m_E which sends a point P to mP:

$$E_m = \ker(m_E : E \to E).$$

Over the field $\mathbb{C}$, the Weierstrass parametrization $E = \mathbb{C}/L$, $L = \mathbb{Z} + \mathbb{Z}z$, gives the isomorphism

$$E_m(\mathbb{C}) \simeq \left(\frac{1}{m}L/L\right) \simeq (\mathbb{Z}/m\mathbb{Z}) \times (\mathbb{Z}/m\mathbb{Z}),$$

i.e. a product of two cyclic groups of order m. The explicit determination of the rational functions defining the map m_E works equally well over any field k whose characteristic is prime to m and when k is algebraically closed gives a similar isomorphism; hence card $E_m(k) = m^2$. When $m = p = \text{char}(k)$, the situation is somewhat more complicated and two distinct cases arise depending on whether the function field of E admits unramified coverings of order p or not. More precisely, suppose that E is given in Legendre normal form

$$E_\lambda : y = x(x - 1)(x - \lambda)$$

and let k be a field of characteristic $p = 2l + 1$. Hasse introduced an invariant of E_λ whose vanishing serves to distinguish between the two cases:

$$H(\lambda) = (-1)^l \sum_{i=0}^{l} \binom{l}{i}^2 \lambda^i.$$

A beautiful theorem of Manin states that the number of points on E_λ which are rational over the finite field $\mathbb{F}_q$, $q = p^f$, satisfies the congruence

$$\text{card } E_\lambda(\mathbb{F}_q) \equiv 1 - H(\lambda)^{(q-1)/(p-1)} \bmod p.$$

If λ is not a root of the equation $H(\lambda) = 0$, then $H(\lambda) \neq 0$ and by choosing the field $\mathbb{F}_q$ suitably we can make the right-hand side of the above congruence be 0, thus forcing $E(\mathbb{F}_q)$ to admit points of order p; indeed, in this case, a more refined analysis shows that the points on $E = E_\lambda$ over $\bar{\mathbb{F}}_Q$ of order p^r satisfy

$$E_{p^r}(\bar{\mathbb{F}}_p) \simeq \mathbb{Z}/p^r\mathbb{Z}.$$

The elliptic curves E_λ defined over a field of characteristic p with $H(\lambda) \neq 0$ are called *ordinary*. When $H(\lambda) = 0$, then $E = E_\lambda$ does not admit points of order p; in this case E_λ is called *supersingular* and the kernel of multiplication by p^r collapses to the origin

$$E_{p^r}(\mathbb{F}_p) = \mathcal{O}.$$

Remark The occurence of $E_{p^r}(\mathbb{F}_p) = \mathcal{O}$ is closely related to the inseparability of the map 'multiplication by p'; the fact that it can occur very often accounts for the possibility of constructing good Goppa codes on the modular curves $X_0(N)$ over the finite field $\mathbb{F}_{p^2}$.

A final useful property of the Hasse invariant is the fact that in the field of formal power series $\mathbb{F}_p[[\lambda]]$ we have

$$H(l) = (-1)^l F(\tfrac{1}{2}, \tfrac{1}{2} : 1 : \lambda),$$

where F is the Gauss hypergeometric function $F = {}_2F_1 \bmod p$; as F satisfies the differential equation

$$\lambda(1 - \lambda)F'' + (1 - 2\lambda)F' - \tfrac{1}{4}F = 0,$$

it follows that each root of $H(\lambda) = 0$ is simple; furthermore, a well-known result from the theory of elliptic curves with complex multiplications shows that $H(\lambda)$ splits completely into a product of linear factors over $\mathbb{F}_{p^2}$.

5.6.5 Igusa's theorem

As a preliminary to Igusa's main result on the modular curves $X_0(N)$, $X_1(N)$, a thorough study of the Galois theory of the field of elliptic functions must precede; this requires some technical constructions with the group of N-division points one of which is the *Weil pairing* e_N. This we now describe. E is taken to be an elliptic curve defined over a field of characteristic 0 or p; $\mathcal{O}$ will denote the origin for the group law on E and N is a positive integer relatively prime to the characteristic of k. Let μ_N be the cyclic group of roots of 1 in the algebraic closure of k. For the construction of the pairing e_N we need the following elementary lemma.

Lemma 5.4 *Let $P_1, \ldots, P_r$ be points on E and $c_1, \ldots, c_r \in \mathbb{Z}$ satisfying*

(i) $\sum\limits_{j=1}^{r} c_j = 0$

(ii) $\sum\limits_{j=1}^{r} c_j \cdot P_j = \mathcal{O}$ *(with addition in the group law for E).*

Then

$$D = \sum_{j=1}^{r} c_j(P_j)$$

is a principal divisor, i.e. $D = (g)$

Proof. As the assertion involves only a finite number of points, we may assume without loss of generality that the exact field of constants k for the function field $k(E)$ of E is algebraically closed; with E as an elliptic curve defined over k and A a divisor on E we let $L(A) = \{f \in k(E) : (f) + A \geq 0\}$. The Riemann–Roch theorem implies $\dim_k L(D + (\mathcal{O})) = 1$; hence there is a function f in the function field $k(E)$, unique up to constant multiples such that $(f) + D + (\mathcal{O}) \geq 0$ and satisfies $(f) + D + (\mathcal{O}) = (x)$ for some $x \in E$. We claim that indeed $x = \mathcal{O}$; this will follow if we note that the sum in the group law of E of the points on the divisor $(f) + D + (\mathcal{O})$ is $\mathcal{O}$. Observe that under the map $P \to (P) - (\mathcal{O})$

$$E \to D_0/D_l,$$

from E to the quotient of the group D_0 of divisors on $k(E)$ of degree 0 by the group D_l of principal divisors, the group law on E is inherited from that on D_0. It is then obvious that the sum of the points on the principal divisor (f) adds up to $\mathcal{O}$. This then proves that $(f) + D + \mathcal{O} = \mathcal{O}$ and hence $D = (f^{-1})$.

To construct the bilinear pairing $e_N(P, Q)$ on $E_N \times E_N$ we fix an N-division point Q on E_N; since the divisor $N(Q) - N(\mathcal{O})$ of degree 0 satisfies the hypothesis of the lemma, it is principal, say

$$N(Q) - N(\mathcal{O}) = (f)$$

for some $f \in k(E)$. Take a point $R \in E$ with $NR = Q$ and apply again the lemma to obtain a function g satisfying

$$\sum_{P \in E_N} (R + P) - \sum_{P \in E_N} (P) = (g).$$

Clearly we have

$$(g^N) = \sum_{P \in E_N} N(R + P) - \sum_{P \in E_N} N(P).$$

We claim this last expression is the same as $(f(Nx))$. This is clear since $R + P$ is a zero of $f(Nx)$ and P is a pole of $f(Nx)$. Thus, up to a constant factor, we have $g(x)^N = f(NX)$ for any $x \in E$. If we take a point $P \in E_N$ then $g(x + P)^N = g(x)^N$. Therefore we have

$$g(x + P) = e_N(P, Q)g(x)$$

for some root of unity e_N in μ_N.

The main properties of the pairing e_N are as follows:

(i) $(P_1 + P_2, Q) = (P_1, Q)(P_2, Q), (P, Q_1 + Q_2) = (P, Q_1)(P, Q_2)$.
(ii) $(P, Q) = (Q, P)^{-1}$.
(iii) If Q is of order N, (P, Q) generates μ_N for some P.
(iv) e_N is non-degenerate, i.e. if $(P, Q) = 1$ for all Q then $P = \mathcal{O}$.
(v) For every automorphism $\sigma \in \mathrm{Gal}(k_s/k)$, k_s the separable closure of the field of definition of E,

$$(P, Q)^\sigma = (\sigma(P), \sigma(Q)).$$

Remark These properties are direct consequences of the method of proof of the above lemma ([78], p. 101).

The map which assigns to a point $P = (x, y)$ on the elliptic curve with model $E: y^2 = 4x^3 - g_2 x - g_3$ its x-coordinate can be given an abstract geometric description as follows. An elliptic curve E defined over a field k can be realized as a double covering of the projective line $\mathbb{P}^1$ ramified at four points. If we identify a point P with its negative in the group law for E, we get a curve which is birationally isomorphic to $\mathbb{P}^1$. Let k be a field of characteristic different from 2 or 3, say $\mathbb{Q}$, and let an elliptic curve be given by

$$E: y^2 = 4x^3 - g_2 x - g_3, \qquad g_i \in k.$$

The group of automorphisms of E can be one of the following:

(i) $\mathrm{Aut}(E) = \pm 1$: this occurs when $j \neq 0, (12)^3$.
(ii) $\mathrm{Aut}(E) = \mu_3: j = (12)^3$, i.e. $g_3 = 0$.
(iii) $\mathrm{Aut}(E) = \mu_4: j = 0$, i.e. $g_2 = 0$.

We now define three functions h_E^i, $1 \leq i \leq 3$ on E, which we denote generically by h:

$$h_E^1((x, y)) = g_2 g_3 x/\Delta, \qquad \Delta = g_2^3 - 27g_3^2,$$

$$h_E^2((x, y)) = g_2^2 x^2/\Delta$$

$$h_E^3((x, y)) = g_3 x^3/\Delta.$$

These functions are clearly defined over k and have the property that if $\eta: E \to E'$ is an isomorphism then $h_E^i = h_{E'}^i \circ \eta$ for $i = 1, 2, 3$. In particular two points P and Q on E satisfy

$$h_E(P) = h_E(Q)$$

if and only if $P = \alpha Q$ for some automorphism $\alpha \in \operatorname{Aut}(E)$.

In describing the Galois Theory of elliptic function fields, we consider a ground field k and a variable element j over k. Representative elements of the isomorphism classes of elliptic curves defined over $k(j)$ with j as their j-invariant are given as follows: if $\operatorname{char}(k) \neq 2, 3$ we take the projective model

$$E: y^2z = 4x^3 - 3^3j(j - 12^3)^{-1}(x + z)z^2,$$

if $\operatorname{char}(k) = 2$ or 3 we take respectively for E the projective models

$$y^2z - xyz = j^{-1}x^3 + jz^3, \qquad y^2z = x^3 - x^2z + j^{-1}z^3.$$

These curves are non-singular and the group law can be defined with reference to the point $\mathcal{O} = (0, 1, 0)$ as origin. If j is allowed to take special values then the resulting specializations E_j remain elliptic curves and have j as their j-invariants provided j is different from $0, 12^3, \infty$.

We now take N to be a positive integer which is not divisible by the characteristic of k and let E_N be the group of N-division points on the elliptic curve E chosen above. Let $k(j, E_N)$ be the finite algebraic extension of $k(j)$ obtained by adjoining to $k(j)$ the coordinates of the points in E_N. Similarly we define $k(j, h(E_N))$ to be the field obtained by adjoining to $k(j)$ the values $h(P)$ of the function h at the points P in E_N. The field $k(j, h(E_N))$ is a Galois extension of $k(j)$ which is intrinsically defined by N, i.e. does not depend on the choice of model for E. If k is extended to its algebraic closure k_a, then $k_a(j, h(E_N))$ is called the *field of modular functions of level N*.

Proposition 5.5 *With notations as above, let j be transcendental over k. Let N be a positive integer not divisible by $\operatorname{char}(k)$, and let E_N be the group of N-division points on E. Then the field $C_N = k(\zeta_N)$ of Nth roots of unity is algebraically closed in $C_N(j, E_N)$. Moreover,*

$$\operatorname{Gal}(C_N(j, E_N)/C_N(j)) \simeq SL_2(\mathbb{Z}/n\mathbb{Z}) \quad and$$

$$\operatorname{Gal}(k(j, E_N)/k(j)) \simeq \left\{ \sigma = \begin{pmatrix} a & b \\ c & d \end{pmatrix} \in GL_2(\mathbb{Z}/n\mathbb{Z}): \sigma(\zeta_N) = \zeta_N^{\det(\sigma)} \right\},$$

$$\operatorname{Gal}(k(j, h(E_N))/k(j)) \simeq SL_2(\mathbb{Z}/N\mathbb{Z})/\{\pm 1_2\}.$$

	char(k) = 2	char(k) = 3	char(k) ≠ 2, 3
P_0	wild $I_0 = A_4, I_1 = V_4$	wild $I_0 = S_3, I_1 = A_3 I_2 - \{1\}$	tame $|I_0| = 3$
P_1	unramified	unramified	tame $|I_0| = 2$
P_∞	tame $|I_0| = N$	tame $|I_0| = N$	tame $|I_0| = N$

Let P_0 (resp. P_1, P_∞) be the points on E corresponding to the discrete valuations of $k_a(j)$ defined by the polynomials j (resp. $j = 12^3, 1/j$). The field $k_a(j, h(E_N))$ as an algebraic extension of $k_a(j)$ ramifies only over the 'points' P_0, P_1, P_∞. Let $I_j(P_0)$ (resp. $I_j(P_1)$, $I_j(P_\infty)$) be the j-th ramification subgroups of the Galois group. The accompanying table gives a complete picture of the possible ramification. The information provided by this table makes it possible to compute, via the Riemann–Hurwitz formula, the genus of all the intermediary function fields.

We now proceed to describe the construction of a non-singular projective model of the field of modular functions of level N in characteristic zero, i.e. $k = \mathbb{Q}$, with the property that its reduction with respect to every prime p not dividing N is a non-singular model for the field of modular functions of level N over $\mathbb{F}_p$. Let $\mathbb{Q}_a$ be the algebraic closure of $\mathbb{Q}$ and $C_N = \mathbb{Q}(\zeta_N)$, $\zeta_N = e^{2\pi i/N}$, the cyclotomic field. Let j be transcendental over $\mathbb{Q}$ and, as before, pick an elliptic curve $E = E_j$ defined over $\mathbb{Q}(j)$ with j as its j-invariant. Let E_N be the group of N-division points and denote by h: $E \to \mathbb{P}^1$ the (Kummer) map defined earlier. The field $\mathbb{Q}_a(j, h(E_N))$ is the field of modular functions with level N-structure in characteristic 0. If the field of algebraic numbers $\mathbb{Q}_a$ is extended to $\mathbb{C}$ we recover the classical field of modular functions of level N. We know that the field $\mathbb{Q}(j, h(E_N))$ does not depend on the choice of model E_j and that C_N is the algebraic closure of $\mathbb{Q}$ in $\mathbb{Q}(j, h(E_N))$. We now fix two generators P, Q of E_N and let j_N be the j-invariant of the elliptic curve $E/(Q)$ obtained from E by dividing out by the cyclic group (Q) generated by Q; j_N is algebraic over $\mathbb{Q}(j)$ and the field

$$K = \mathbb{Q}(j, j_N, h(P))$$

is an intrinsically defined algebraic extension of $Q(j)$ which depends only on N and not on the choice of model E_j or on the choice of generators P and Q for E_N. From the Galois theory described in Proposition 5.5, we see that $\mathbb{Q}(j, h(E_N))$ and $\mathbb{Q}(j)$ are respectively the compositum and the intersection of K and $C_N(j)$.

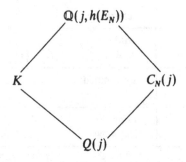

Let p be a prime number and extend the p-adic valuation ord_p of $\mathbb{Q}$ to the field of rational functions $\mathbb{Q}(j)$ as usual: write any rational function in $\mathbb{Q}(j)$ as a rational number times the quotient of two polynomials in $\mathbb{Z}[j]$, each with coefficients free of common divisors and assign to it the order of the rational number. Denote by r the corresponding valuation ring of $\mathbb{Q}(j)$:

$$r = \{A(j) \in \mathbb{Q}(j) \colon \mathrm{ord}_p A(j) \geq 0\}.$$

Let $\bar{j}$ be transcendental over the finite $\mathbb{F}_p$; then r is the localization of the ring $\mathbb{Z}[j]$ with respect to the prime ideal which defines the reduction map $\mathbb{Z}[j] \to \mathbb{F}_p[j]$.

The following two lemmas, which we state without proof, are essential ingredients in Igusa's construction of the smooth model for the field K. First recall that the integral closure of a ring A contained in a field B is the ring of all elements in B which are roots of monic polynomials in $A[X]$.

Lemma 5.5 *If p does not divide the level N, the integral closure R of r in K is an unramified valuation ring.*

To state the second result we need some notation. Let S be the integral closure of $\mathbb{Z}[j]$ in K. By picking a basis $\theta_1, \ldots, \theta_\mu$ of K over $\mathbb{Q}(j)$, with discriminant d and observing that as a $\mathbb{Z}[j]$-module S is contained in $\mathbb{Z}[j]$ $[\theta_1/d, \ldots, \theta_\mu/d]$, we obtain from the Noetherian property of $\mathbb{Z}[j]$ that S is finitely generated over $\mathbb{Z}[j]$. Also let T be the integral closure of $\mathbb{Z}[1/j]$ in K; for similar reasons as above T is a finitely generated module over $\mathbb{Z}[1/j]$. As $\mathbb{Z}[j]$ and $\mathbb{Z}[1/j]$ are contained in the valuation ring r, both S and T are contained in the integral closure of r in K. If $p \nmid N$, Lemma 5.5 guarantees that R is a local ring with maximal ideal pR. In fact, as is easily verified using a theorem of Krull, $pR \cap S = pS$. Similarly we get $pR \cap T = pT$; thus pS and pT are prime ideals respectively of S and T. Let k be the residue class field of R, i.e. $k = R/pR$. Let $\bar{S}$ and $\bar{T}$ be the images of S and T under the residue homomorphism $R \to k = R/pR$, i.e. $\bar{S} = S/pS$ and $\bar{T} = T/pT$. The residue class field k is exactly the field of quotients of $\bar{S}$ and $\bar{T}$; the integral domains $\bar{S}$ and $\bar{T}$ are integrally closed in k.

Lemma 5.6 *The residue class field $k = R/pR$ is a regular extension of $\mathbb{F}_p$, that is $\mathbb{F}_p$ is algebraically closed in k and k is a separable extension. Furthermore the compositum of k and the algebraic closure of $\mathbb{F}_p$ is the field of modular functions of level N in characteristic p.*

The so-called Kroneckerian model of the field of modular functions of level N is the union $\mathcal{M}$ of all discrete valuation rings of S and T. The union $\mathcal{D}$ of all discrete valuation rings of $\mathbb{Z}[j]$ and $\mathbb{Z}[1/j]$ is the universal projective line and $\mathcal{M}$ is in some sense 'the derived normal model' of $\mathcal{D}$ in K. If the discrete valuation rings of $\mathcal{M}$ are classified according to the characteristic of the corresponding residue class fields, then the model $\mathcal{M}$ is the disjoint union of local models, each parametrized by the primes of $\mathbb{Q}$:

$$\mathcal{M} = \mathcal{M}_0 + \sum_p \mathcal{M}_p,$$

where $\mathcal{M}_0$ is the model corresponding to characteristic zero. The main idea of Igusa is to construct a projective model of $\mathcal{M}_0$ such that $\mathcal{M}_p$ is the union of the local rings of points on the reduction of $\mathcal{M}_0$ modulo p.

Let m be a positive integer and let S_m be the set of elements f in S such that f/j^m is integral over $\mathbb{Q}[1/j]$. Put $T_m = T \cap (1/j)^m S$. Clearly S_m contains $1, j, \ldots, j^m$ and T_m contains $1, 1/j, \ldots, (1/j)^m$. If we take m sufficiently large we obtain $\mathbb{Z}[S_m] = S$ and $\mathbb{Z}[T_m] = T$. Applying the reduction map $R \to k = R/pR$, we get $\mathbb{F}_p[\bar{S}_m] = \bar{S}$ and $\mathbb{F}_p[\bar{T}_m] = \bar{T}$. Let $f_0, f_1, \ldots, f_n$ be a basis for the free $\mathbb{Z}$-module S_m. Let C be the abstract curve corresponding to the field K, i.e. the collection of closed points corresponding to discrete valuation rings of K, and let

$$C \xrightarrow{\pi} \mathbb{P}^M$$

be the map which sends a point $x \to (f_0(x), \ldots, f_M(x))$. Once m is fixed, the image of π is a curve, also denoted by C, which is uniquely determined up to a projective transformation induced by a matrix in $SL_{M+1}(\mathbb{Z})$. A simple argument shows that it is possible to normalize C by taking as a basis $f_0 = 1, f_1 = j, \ldots, f_m = j^m$.

The ring theoretic properties of S and T can now be translated into geometric properties of C. First of all C is non-singular. In fact the projection of C on the hyperplane at infinity in $\mathbb{P}^n$ corresponding to the first coordinate is an affine curve with coordinate ring $S \otimes_{\mathbb{Z}} \mathbb{Q}$, which is the integral closure of $\mathbb{Q}[j]$ in K. This affine curve is normal over $\mathbb{Q}$ and hence non-singular. Similarly the projection of C on the hyperplane at infinity in $\mathbb{P}^M$ corresponding to the $(M + 1)$ coordinate is an affine curve with coordinate ring $T \otimes_{\mathbb{Z}} \mathbb{Q}$ which is the integral closure of $\mathbb{Q}[1/j]$ in K. The resulting affine curve is also normal over $\mathbb{Q}$ and hence non-singular. These two affine curves, which provide a complete open cover for C, are fixed in the discussion which follows. The crucial property of C is the claim that for a prime p which does not divide N, the reduction C_p of C with respect to p remains non-singular. As indicated in Lemma 5.6, pS is a prime ideal of S and the field of quotients k of $\bar{S} = S/pS$ is regular over $\mathbb{F}_p$. This implies that the reduction of one open set of C with respect to p is an irreducible affine curve over $\mathbb{F}_p$. Now, since $\mathbb{F}_p[\bar{S}_m] = \bar{S}$ and $\bar{S}$ is integrally closed in k, this curve is normal over $\mathbb{F}_p$ and hence non-singular. Similarly the reduction of the other open set of C with respect to p is non-singular. These two affine curves determine a complete open cover of C_p, and hence the curve C_p is non-singular. The following is then the fundamental theorem in the theory of elliptic modular functions.

Theorem 5.9 (Igusa) *There exists a non-singular projective model C of the field K over $\mathbb{Q}$ such that the local model $\mathcal{M}_p$ is the union of discrete valuation rings corresponding to the closed points on its reduction C_p with respect to p. If $p \nmid N$ then C_p is a non-singular projective model of the field $k = R/pR$ over $\mathbb{F}_p$.*

The same reasoning outlined above provides non-singular projective models for any function field intermediate between K and $\mathbb{Q}(j)$ with similar properties under reduction modulo p. In particular this applies to the field $K_0 = \mathbb{Q}(j, j_N)$. We state this as a corollary.

Corollary 5.9.1 *The field $\mathbb{Q}(j, j_N)$ has a non-singular projective model C defined over $\mathbb{Q}$ with good reduction modulo primes p which do not divide N.*

Definition 5.18 *The smooth projective model C defined over $\mathbb{Q}$, whose existence is ascertained by the corollary above is the curve associated to the Riemann surface*
$$X_0(N)_{\mathbb{C}} = H/\Gamma_0(N) \cup \mathbb{P}^1(\mathbb{Q})/\Gamma_0(N).$$

In the following we denote this curve by $X_0(N)$ and think of $X_0(N)_{\mathbb{C}}$ as corresponding to the complex points on C.

Remark The deeper properties of the curve C have been studied by considering C as a moduli variety for elliptic curves with certain types of level N-structures. Let us recall briefly what this means. If E is an elliptic curve defined over an algebraically closed field k of characteristic p not dividing N, then the group of N-division points on E is endowed with a non-degenerate bilinear pairing

$$e_N \colon E_N \times E_N \to \mu_N.$$

A level N-structure on the elliptic curve E is simply an isomorphism

$$\alpha \colon E_N \simeq (\mathbb{Z}/n\mathbb{Z}) \times (\mathbb{Z}/n\mathbb{Z}).$$

If such a level N-structure α is given, let $\zeta(\alpha)$ denote the primitive N-th root of unity

$$\zeta(\alpha) = e_N(\alpha^{-1}(1,0), \alpha^{-1}(0,1)).$$

Suppose now that Γ_N denotes the principal congruence subgroup

$$\Gamma(N) = \left\{ \begin{pmatrix} a & b \\ c & d \end{pmatrix} \in \Gamma \colon \begin{pmatrix} a & b \\ c & d \end{pmatrix} \equiv \pm 1_2 \bmod N \right\}.$$

The (incomplete) Riemann surface $H/\Gamma(N)$ is the moduli variety over $\mathbb{C}$ which parametrizes isomorphism classes of elliptic curves defined over $\mathbb{C}$ endowed with a level N-structure α satisfying $\zeta(\alpha) = e^{2\pi i/N}$.

A compact Riemann surface is obtained from $H/\Gamma(N)$ by adding the orbits $\mathbb{P}^1(\mathbb{Q})/\Gamma(N)$ with appropriate neighborhoods; it also has an interesting interpretation as a moduli variety. The function field of this moduli variety has a smooth model M_N defined over $\mathbb{Z}$ and for its complex points we have

$$M_N(\mathbb{C}) = H/\Gamma(N) \cup \mathbb{P}^1(\mathbb{Q})/\Gamma(N).$$

The consideration of M_N as a moduli scheme, in the terminology of modern algebraic geometry, requires the consideration of generalized elliptic curves with level N-structures. Such objects are defined as follows. Let S be a scheme, say a ring like $\mathbb{Z}$ or the ring of integers of a number field. A generalized elliptic curve over the base S is a curve of arithmetic genus 1 over S together with a proper flat morphism of finite presentation as shown in Figure 5.9, endowed with a section e which is continuous on the complement of the singular locus of π and where the geometric fibers are irreducible reduced curves of arithmetic genus 1. The geometric fibers of π are of the following type:

Figure 5.9

(i) An elliptic curve, i.e. an absolutely irreducible smooth projective curve of genus 1.

(ii) A projective line on which two points have been identified, e.g. a plane cubic in $\mathbb{P}^2$ with an ordinary double point.

(iii) A projective line on which two infinitely near points have been identified, e.g. a plane cubic in $\mathbb{P}^2$ with a cusp.

The main result in the theory is the existence of a moduli scheme for the space of isomorphism classes of elliptic curves with level N-structure. The crudest example of such a scheme is provided by

$$\operatorname{Spec} \mathbb{Z}[\tfrac{1}{2}, \tfrac{1}{3}][g_2, g_3],$$

which serves as a moduli for curves of genus 1 endowed with a non-zero invariant differential. The universal curve is $y^2 = 4x^3 - g_2 x - g_3$ and the invariant differential is $\omega = dx/y$. We shall not pursue these remarks here and recommend the interested reader consult the fundamental papers in the theory (Deligne & Rappoport [12], Katz [43], and Mazur [58]).

5.6.6 The modular equation

Let $E = \mathbb{C}/L$, $L = \mathbb{Z} + \mathbb{Z}z$, $z \in H$. As the function $j = j(z)$ is an invariant of the isomorphism class of the lattice L, it has a Fourier expansion which is given by

$$j(z) = (12)^3 g_2(L)^3/(g_2(L)^3 - 27 g_3(L)^2)$$

$$= \left\{1 + 240 \sum_{n=1}^{\infty} \sigma_3(n) t^n\right\}^3 \Big/ t \prod_{n=1}^{\infty} (1 - t^n)^{24}, \qquad t = \exp 2\pi i z,$$

$$= t^{-1}\left\{1 + \sum_{n=1}^{\infty} c_n t^n\right\}, \qquad c_n \in \mathbb{Z}.$$

A basis fact in the theory of modular forms ascertains that the ring of holomorphic modular forms for the unimodular group $\Gamma = SL_2(\mathbb{Z})/\{\pm 1_2\}$ is generated by the Eisenstein series $g_2(L) = 60 E_4(z)$ and $g_3(L) = 140 E_6(z)$ subject to the single relation $\Delta = g_2(L)^3 - 27 g_3(L)^2$. A consequence of this is that any modular function (i.e. a holomorphic function on H invariant under Γ) with at most a pole at infinity is a polynomial in $j(z)$. We need the following version of the so-called *t-expansion principle*.

Lemma 5.7 *Let* $t = \exp 2\pi i z$. *Suppose that the equality*

$$\sum_{k=0}^{m} a_k j(z)^k = \sum_{n \geq n_0} b_n t^n$$

holds for all $z \in H$, with constants a_k, b_n in $\mathbb{C}$. Then the a_k belong to the ring generated by the b_n over $\mathbb{Z}$.

Proof. Substitute the expansion $t^{-1}\{1 + \sum_{n=1}^{\infty} c_n t^n\}$ for $j(z)$ in the polynomial $\sum_{k=0}^{m} a_k j(z)^k$:

$$a_m\left(1 + \sum_{n=1}^{\infty} c_n t^n/t\right)^m + a_{m-1}\left(1 + \sum_{n=1}^{\infty} c_n t^n/t\right)^{m-1}$$

$$+ a_{m-2}\left(1 + \sum_{n-1}^{\infty} c_n t^n/t\right)^{m-2} + \cdots = \sum_{n \geq n_0} b_n t^n.$$

Using the multinomial theorem and equating coefficients we get

coeff. of t^{-m} is $a_m = b_{-m}$

coeff. of $t^{-(m-1)}$ is $a_m m c_1 + a_{m-1} = b_{1-m}$

coeff. of $t^{-(m-2)}$ is $a_m \dfrac{m(m-1)}{2} c_2 + a_{m-1}(m-1)c_1 + a_{m-2} = b_{2-m}$

and so forth. Since $c_n \in \mathbb{Z}$, the a_n's are representable as $\mathbb{Z}$-linear combinations of the b_n's. This proves the lemma.

The elements of the set

$$M_n = \left\{\alpha = \begin{pmatrix} a & b \\ c & d \end{pmatrix} \in M_2(\mathbb{Z}): \det \alpha = n, (a,b,c,d) = 1\right\}$$

are called primitive matrices. To a pair consisting of a primitive matrix $\alpha = \begin{pmatrix} a & b \\ c & d \end{pmatrix}$ in M_n and a lattice $L = \mathbb{Z}z + \mathbb{Z}$ there is associated another lattice $L' = \mathbb{Z}(az+b) + \mathbb{Z}(cz+d)$ of index n in L. By the elementary divisor theorem there is a basis ω_1, ω_2 for L such that $L = \mathbb{Z}\omega_1 + \mathbb{Z}\omega_2$ and a basis ω_1', ω_2' for L' such that $L = \mathbb{Z}\omega_1' + \mathbb{Z}\omega_2'$ and $\omega_1' = e_1\omega_1, \omega_2' = e_2\omega_2$ and $e_1 | e_2$. Now since $(a,b,c,d) = 1$, we have $e_1 = 1$, $e_2 = n$. In particular L/L' is cyclic of order n. Thus there are matrices γ, γ' such that

$$\gamma \begin{pmatrix} a & b \\ c & d \end{pmatrix} \gamma' = \begin{pmatrix} 1 & 0 \\ 0 & n \end{pmatrix}.$$

From the identity $\begin{pmatrix} 0 & -1 \\ 1 & 0 \end{pmatrix}\begin{pmatrix} 1 & 0 \\ 0 & n \end{pmatrix}\begin{pmatrix} 0 & 1 \\ -1 & 0 \end{pmatrix} = \begin{pmatrix} n & 0 \\ 0 & 1 \end{pmatrix}$ we obtain that $M_n = \Gamma \begin{pmatrix} n & 0 \\ 0 & 1 \end{pmatrix} \Gamma$. If $\alpha = \begin{pmatrix} a & b \\ c & d \end{pmatrix} \in \Gamma \begin{pmatrix} n & 0 \\ 0 & 1 \end{pmatrix} \Gamma$, there is a unimodular matrix $\gamma = \begin{pmatrix} x & y \\ z & w \end{pmatrix} \in \Gamma$ such that $za + wc = 0$ and $xw - yz = 1$; thus

$\gamma\alpha = \begin{pmatrix} a_1 & b_1 \\ 0 & d_0 \end{pmatrix}$. Suppose now that α is in triangular form, say $\alpha = \begin{pmatrix} a & b \\ 0 & d \end{pmatrix}$,

then, since $\begin{pmatrix} 1 & k \\ 0 & 1 \end{pmatrix}\begin{pmatrix} a & b \\ 0 & d \end{pmatrix} = \begin{pmatrix} a & b + kd \\ 0 & d \end{pmatrix}$, we obtain a disjoint decomposition

$$\Gamma\begin{pmatrix} n & 0 \\ 0 & 1 \end{pmatrix}\Gamma = \bigcup_{\alpha \in A} \Gamma\alpha,$$

where the coset representatives $\alpha \in A$ satisfy

$$\alpha = \begin{pmatrix} a & b \\ 0 & d \end{pmatrix}, \quad d > 0, \quad ad = n, \quad 0 \le b < d, \quad (a,b,d) = 1.$$

A count for the number $\psi(n)$ of coset representatives is obtained by observing that for each divisor d of n there are $(d/e)\varphi(e)$, $e = (n/d, d)$, possible values for b. Thus

$$\psi(n) = \sum_{d|n} \frac{d}{e}\varphi(e).$$

Now, elementary considerations show that $\psi(nn') = \psi(n)\psi(n')$ if n and n' are relatively prime; also we have $\psi(p^r) = p^r(1 + 1/p)$ and therefore

$$\psi(n) = n \prod_{p|n} (1 + 1/p).$$

If $n = p$ is a prime we have

$$A = \left\{ \begin{pmatrix} p & 0 \\ 0 & 1 \end{pmatrix}, \quad \begin{pmatrix} 1 & i \\ 0 & p \end{pmatrix}, \quad 0 \le i \le p - 1 \right\}.$$

For $\alpha = \begin{pmatrix} a & b \\ c & d \end{pmatrix}$ we introduce the notation $j \circ \alpha = j\left(\dfrac{az + b}{cz + d}\right)$. Fix an integer N and define the polynomial in x

$$\prod_{\alpha \in A} (x - j \circ \alpha) = \sum_{m=0}^{\psi(N)} s_m x^m,$$

where A is a set of representatives for the primitive matrices of determinant N and the coefficients s_m are the elementary symmetric functions in the $j \circ \alpha$. As we have $\bigcup_{\alpha \in A} \Gamma\alpha \cdot \gamma = \bigcup_{\alpha \in A} \Gamma\alpha$ every $\gamma \in \Gamma$, composing $j \circ \alpha$ with γ induces a permutation of the set $\{j \circ \alpha\}_{\alpha \in A}$. This implies that the elementary symmetric functions s_m satisfy $s_m \circ \gamma = s_m$ and are therefore holomorphic functions in H invariant under Γ and with at most a pole at infinity. Therefore they are polynomials in j:

$$s_m = s_m(j).$$

Furthermore, the s_m have power series expansions in $t^{1/N}$ with coefficients in $\mathbb{Z}[\zeta_N]$, $\zeta_N = \exp(2\pi i/N)$. In fact for $\alpha = \begin{pmatrix} a & b \\ 0 & d \end{pmatrix} \in A$, we have

$$j \circ \alpha = j\left(\frac{az+b}{d}\right) = \zeta_d^{-b} t^{-a/d}\left(1 + \sum_{n=1}^{\infty} c_n \zeta_d^{mb} t^{ma/d}\right),$$

$\zeta_d = e^{2\pi i/d}$ and the coefficients of $t^{a/d}$ clearly lie in $\mathbb{Z}[\zeta_N]$. Under the canonical isomorphism $\mathrm{Gal}(\mathbb{Q}(\zeta_N)/\mathbb{Q}) \simeq (\mathbb{Z}/\mathbb{Z}N)^\times$ let the integer s (mod N) correspond to the automorphism σ, i.e. $\sigma(\zeta_N) = \zeta_N^s$. The action of σ on the coefficients of $j \circ \alpha$ produces a $j \circ \beta$ with some

$$\beta = \begin{pmatrix} a & b' \\ 0 & d \end{pmatrix} \in A, \qquad \text{and} \qquad b' \equiv sb \bmod d.$$

The map $\sigma^*: A \to A$ given by $\sigma^*\begin{pmatrix} a & b \\ 0 & d \end{pmatrix} = \begin{pmatrix} a & b' \\ 9 & d \end{pmatrix}$ with $b' \equiv sb \bmod d$ induces a permutation of the set A; therefore the elementary symmetric functions in the $j \circ \alpha$ are fixed under the action of σ. Hence, by the t-expansion principle, $s_m = s_m(j)$ is a polynomial in $\mathbb{Z}[j]$. Thus

$$H_N(x,j) = \prod_{\alpha \in A} (x - j \circ \alpha) = \sum_{m=0}^{\psi(N)} s_m(j) x^m$$

is a polynomial in $\mathbb{Z}[x,j]$.

If $\xi = \begin{pmatrix} a & b \\ c & d \end{pmatrix} \in \mathrm{GL}_2(\mathbb{Q})$ with $\det \xi > 0$, then $j \circ \xi$ is integral over $\mathbb{Z}[j]$, i.e. satisfies a monic polynomial with coefficients in $\mathbb{Z}[j]$. In fact, multiplying ξ by a suitable integer, an operation which does not change the value of $j((az+b)/(cz+d))$, we may assume ξ is a primitive element. If $\det \xi = N > 1$ then $\xi \in \Gamma \begin{pmatrix} N & 0 \\ 0 & 1 \end{pmatrix}$ and $\Gamma \xi = \Gamma \alpha$ for some $\alpha \in A$. Therefore $j \circ \xi = j \circ \alpha$ and clearly $H_N(j \circ \xi, j) = 0$. In particular we see that

$$H_N(j(Nz), j(z)) = 0.$$

The equation $H_N(x, y) = 0$ provides an affine model for the Riemann surface

$$X_0(N)_{\mathbb{C}} = H/\Gamma_0(N) \cup \mathbb{P}^1(\mathbb{Q})/\Gamma_0(N),$$

which unfortunately is highly singular. As indicated earlier the open set $Y_0(N)_{\mathbb{C}} = H/\Gamma_0(N)$ provides the moduli space for the isomorphism classes of pairs (E, C) consisting of an elliptic curve E defined over $\mathbb{C}$ together with a cyclic subgroup of order N.

The polynomial $H_N(j) = H_N(j,j)$ is called the modular equation; it has important properties that make it quite useful in the theory of elliptic curves with complex multiplications. Here we mention only the following proposition.

Proposition 5.6 *If N is not a square, the coefficient of the highest power of j in $H_N(j)$ is* 1.

Proof. As N is not a square, all coset representatives $\alpha = \begin{pmatrix} a & b \\ c & d \end{pmatrix}$ in A satisfy $a/d \neq 1$ and hence

$$j(z) = j((az + b)/c) = \zeta t^{-e} + \text{(higher powers of } t),$$

where $e = \max(1, a/d)$ and ζ is a unit in $\mathbb{Z}[\zeta_N]$. From the t-expansion principle we obtain that the leading term in the t-expansion of the product

$$\prod (j(z) - j((az + b)/d))$$

is a unit in $\mathbb{Z}$ and hence $= \pm 1$ this proves the result.

An argument similar to that above, using some simple properties of the binomial coefficients, leads to the following fundamental property of the equation $H_N(x, y)$.

Proposition 5.7 (Kronecker's congruence relation) *For p a prime number we have*

$$H_p(x, y) \equiv (x^p - y)(x - y^p) \bmod p.$$

Remark A hint of the rich structure of the curve $X_0(p)$ in characteristic p is given by Kronecker's congruence relation. In fact over the algebraic closure $\bar{\mathbb{F}}_p$,

$$X_0(p)(\bar{\mathbb{F}}_p) = \mathbb{P}^1(\bar{\mathbb{F}}_p) \cup \mathbb{P}^1(\bar{\mathbb{F}}_p),$$

i.e. the union of two projective lines. The singularities are ordinary double points, all defined over $\mathbb{F}_{p^2}$. A more refined analysis using the moduli interpretation of $X_0(p)$ reveals that the singularities actually occur at the points corresponding to supersingular elliptic curves (Figure 5.10).

Example There are formulas for calculating the coefficients c_n in the expansion $j(z) = t^{-1} + \sum_{n=0}^{\infty} c_n t^n$ and numerical tables are available. The method we have used to show that the coefficients of $H_N(x, j)$ are in $\mathbb{Z}[j]$ can also be used for actual numerical calculation. Except for small values of N, the size of the c_n makes this task difficult and expensive even for a present day supercomputer. On the other hand the same algorithm can be carried out modulo a prime l, say for l not dividing N. The resulting equation for

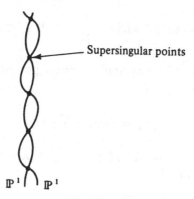

Figure 5.10

$H_N(x, y)$ over $\mathbb{F}_l$ will give an affine model for $X_0(N)$ in characteristic l. For $N = 2$, the following equation for $H_2(x, y)$ is well known:

$$x^3 + y^3 - x^2 y^2 + 1488xy(xy) - 162\,000(x^2 + y^2) + 40\,773\,375xy$$
$$+ 8\,748\,000\,000(x + y) - 157\,464\,000\,000\,000 = 0.$$

5.6.7 The congruence formula

On the curve $X_0(N)$ over the field $\mathbb{Q}$ the space of differentials $\Omega_0^{\mathrm{cusp}}(N)$ of the first kind which vanish about the cusps is isomorphic to the space $S_2^0(\Gamma_0(N))$ of holomorphic cusp forms defined on H of weight 2 for the group $\Gamma_0(N)$. Recall that such forms are defined by functions $f: H \to \mathbb{C}$ which are holomorphic, vanish about all cusps, and under the action of $\Gamma_0(N)$ satisfy

$$f\left(\frac{az + b}{cz + d}\right) = (cz + d)^2 f(z);$$

the isomorphism is given by $f(z) \to \omega = f(z)\,dz$. Let $t = \exp(2\pi i z)$ be a local uniformizing parameter about the cusp at infinity and let

$$f = \sum_{n=1}^{\infty} a(n) t^n$$

be the expansion of a cusp form in powers of t. If p is a prime not dividing N, we define the Hecke operator T_p by the formula

$$T_p \circ f = \sum_{n=1}^{\infty} a(n) t^{pn} + \sum_{n=1}^{\infty} a(pn) t^n.$$

The proof that T_p leaves the space $S_2^0(\Gamma_0(N))$, and hence $\Omega_0^{\mathrm{cusp}}(N)$, invariant is based on the fact that the set

$$v_\infty(z) = pz, \qquad v_i(z) = \frac{z+i}{p}, \qquad 0 \le i \le p-1$$

forms a complete set of representatives for the primitive matrices $\begin{pmatrix} a & b \\ c & d \end{pmatrix}$ with determinant p and

$$T_p \circ f = f \circ v_\infty + \sum_i f \circ v_i.$$

It is possible to define operators for the higher powers p^k and show that they satisfy the relation for $k \ge 2$

$$T_{p^k} = T_{p^{k-1}} \circ T_p - pT_{p^{k-2}}, \qquad T_i = \mathrm{Id}.$$

Iqusa's theorem on the model of $X_0(N)$ shows that for a prime p, $p \nmid N$, the space of differentials of the first kind on $X_0(N)(\mathbb{F}_p)$ is obtained by reducing modulo p the corresponding space for $X_0(N)_\mathbb{Q}$. Iqusa's theorem in fact provides the existence of a basis for $\Omega_0^{\mathrm{cusp}}(N)$ whose t-expansions have rational coefficients in $\mathbb{Q}$ with denominators in $\mathbb{Q} \cap \mathbb{Z}_p$; the reduction of these forms provide a basis for the differentials of the first kind on $X_0(N)_{\mathbb{F}_p}$. Using this procedure it makes sense to reduce the Hecke operator T_p modulo p. The resulting operator turns out to be the Frobenius operator on $X_0(N)_{\mathbb{F}_p}$ and its transpose acting on the space of differentials of the first kind. This is basically the content of the congruence relation; a proof of this result, which is originally due to Eichler, can be based on the congruence relation $H_p(x, y) \equiv (x^p - y)(x - y^p) \bmod p$. As the trace of Frobenius is related to the number of points on $X_0(N)(\mathbb{F}_p)$, we now state the result in the following form, which is quite suitable for applications to Goppa codes.

Theorem 5.10 *Let p be a prime not dividing N. Then we have*

$$\mathrm{card}\, X_0(N)(\mathbb{F}_p) = p + 1 - T(p),$$

where $T(p) = \mathrm{trace}(T_p)$ acting on the space of cusp forms of weight 2 for $\Gamma_0(N)$.

We would like to add a few words about the proof of this important theorem. Our brief discussion is intended to suggest only how the main line of the argument runs and we invite the interested reader to consult the very elucidating papers of Eichler [23] and Shimura [78]. To begin with we start with a prime p which does not divide the level N. As implied by Iqusa's theorem, the physical appearance of the curve $X_0(N)$ is the same over the complex numbers as over the algebraic closure of $\mathbb{F}_p$. The congruence relation is then an equality between the Frobenius correspondence acting on $X_0(N)(\bar{\mathbb{F}}_p)$ and the Hecke correspondence acting on $X_0(N)(\mathbb{C})$. More precisely, in characteristic p, the Frobenius correspondence is defined on points as follows: A point P on $X_0(N)(\bar{\mathbb{F}}_p)$ corresponds to an isomorphism class of pairs (E, C) consisting of an elliptic curve E together with a cyclic subgroup C of order N. Now the isomorphism class of E is determined by the j-invariant; hence

$$P \leftrightarrow E \leftrightarrow j = j(E).$$

The Frobenius correspondence takes P into the point Q associated to the elliptic curve $E^{(p)}$ obtained from E by raising the coefficients to the p-th power; that is to say, $E^{(p)}$ is an elliptic curve in the isomorphism class determined by the j-invariant j^p; thus

$$\mathrm{Frob}(P) = Q \leftrightarrow E^{(p)} \leftrightarrow j^p.$$

We also need Frob*, which is the transpose of the correspondence Frob. This is defined as

$$\mathrm{Frob}(P) = P_1 + P_2 + \cdots + P_p,$$

where

$$P_i \leftrightarrow E_i \leftrightarrow j_i = j(E_i)$$

and the isomorphism class of E_i is that characterized by the invariant j_i satisfying

$$j_i^p = j,$$

or equivalently E_i satisfies $E_i^{(p)} = E$. The sum Frob + Frob* defines a correspondence on the points P of the curve $X_0(N)(\bar{\mathbb{F}}_p)$ by setting

$$(\mathrm{Frob} + \mathrm{Frob}^*)(P) = Q + P_1 + \cdots + P_p.$$

We can extend this correspondence by additivity to the group of divisor classes. We now pass to the situation in characteristic zero and suppose that the point P in $X_0(N)(\bar{\mathbb{F}}_p)$ comes from reduction modulo p of a point P (which by abuse of notation we also denote by the same symbol) in $X_0(N)_\mathbb{C}$. Recall that such a point P is associated to an isomorphism class of elliptic curves, of which a representative can be expressed as $E(z) = \mathbb{C}/(\mathbb{Z} + \mathbb{Z}z)$, with z a complex number in H. Such an isomorphism class is characterized by the value of the j-invariant

$$j(z) = t^{-1} + \sum_{n=0}^{\infty} c_n t^n, \qquad t = \exp(2\pi i z).$$

Thus we have

$$P \leftrightarrow E(z) = \mathbb{C}/(\mathbb{Z} + \mathbb{Z}z) \leftrightarrow j = j(z).$$

Since the coefficients c_n are ordinary integers, a simple extension of Fermat's little theorem to the ring of formal power series in t over $\mathbb{Z}$ gives the congruence

$$j(z)^p \equiv \widetilde{j(pz)} \bmod p.$$

With the invariant $j(pz)$ is associated the isomorphism class of the elliptic curve $E(pz) = \mathbb{C}/(\mathbb{Z} + \mathbb{Z}pz)$. This then suggests the congruence

$$E^{(p)} \equiv \tilde{E}(pz) \bmod p,$$

where the tilde means that the equation for $E(pz)$ has been reduced modulo p. This is not always possible and requires that the j-invariant of E does not lie in $\mathbb{F}_{p^2}$; we shall not stop here to explain this difficulty. Again a straightforward application of Fermat's little theorem gives the congruence

$$j\left(\frac{z+i}{p}\right)^p \equiv j(z) \bmod p, \qquad 0 \le i \le p - 1;$$

therefore $j_i = j((z + i)/p)$ can be considered as the j-invariant of the iso-morphism class of the elliptic curve $E((z + i)/p) = \mathbb{C}/(\mathbb{Z} + \mathbb{Z}(z + i)/p)$. This suggests the congruence

$$E_i \equiv \tilde{E}\left(\frac{z+i}{p}\right) \bmod p.$$

The map

$$T_p(z) = (pz) + \sum_{i=0}^{p-1}\left(\frac{z+i}{p}\right)$$

acting on the points of $X_0(N)_\mathbb{C}$ is just the Hecke correspondence acting at the level of homology and by transport of structure its action on the space of differentials of the first kind on $X_0(N)_\mathbb{C}$ is simply the ordinary Hecke operator on the space of cusp forms of weight 2 for $\Gamma_0(N)$: for a form

$$f = \sum_{n=1}^{\infty} a(n)t^n,$$

$$f \circ T_p = \sum_{n=1}^{\infty} a(n)t^{pn} + \sum_{n=1}^{\infty} a(pn)t^n.$$

The foregoing discussion makes the relation

$$\text{Frob} + \text{Frob}^* = \tilde{T}_p$$

quite plausible. To obtain the actual equality claimed in the theorem it is necessary first to represent the correspondences Frob, Frob* and $\tilde{T}_p$ on a vector space of characteristic zero and then to take traces, using the fact that the fixed points of Frobenius acting on $X_0(N)(\overline{\mathbb{F}}_p)$ correspond to those which have coordinates in $\mathbb{F}_p$. Such a representation of the Frobenius and Hecke correspondence on the same vector space over a field of characteristic zero leads, via the congruence relation, to the identity

$$Z(X_0(N), x) = \frac{H_p(x)}{(1 - x)(1 - px)}.$$

Equating the coefficients of x after logarithmic differentiation gives the identity

$$N_1 = p + 1 - T(p)$$

$$= p + 1 - \alpha_1 - \alpha_2 - \cdots - \alpha_{2g},$$

where $T(p)$ is the trace of T_p acting on $S_2^0(\Gamma_0(N))$, and $\alpha_1, \ldots, \alpha_{2g}$ are the reciprocal roots of the polynomial $H_p(x)$. More generally, for an arbitrary positive integer n we have

$$N_n = p^n + 1 - \alpha_1^n - \cdots - \alpha_{2g}^n.$$

The space $S_2^0(\Gamma_0(N))$ is of dimension $g_0 = $ genus of $X_0(N)$; let $D(T_p)$ be a matrix representation of T_p in $S_2^0(\Gamma_0(N))$. The p-th *Hecke polynomial* is now defined by

$$H_p(x) = \det(I_g - xD(T_p) + px^2 I_g)$$

$$= (1 - \alpha_1 x) \ldots (1 - \alpha_{2g} x).$$

Let N_n denote the number of points P on $X_0(N)$ which are of degree 1 over $\mathbb{F}_{p^n}$; recall that these are the points P on $X_0(N)(\overline{\mathbb{F}}_p)$ whose residue class fields $\mathbb{F}(P)$ has degrees over $\mathbb{F}_p$ which are divisors of n. The zeta function of $X_0(N)$ over $\mathbb{F}_p$ is defined by

$$Z(X_0(N), x) = \exp \sum_{n=1}^{\infty} N_n \frac{c^n}{n}.$$

Let $a_1, \ldots, a_g$ be the eigenvalues of T_p and put

$$1 - a_i x + px^2 = (1 - \alpha_i x)(1 - \alpha_i' x), \qquad 1 \le i \le g.$$

Then we have

$$H_p(x) = \prod_{1 \le i \le g} (1 - \alpha_i x)(1 - \alpha_i' x)$$

$$= \exp \left\{ - \sum_{m=1}^{\infty} \left(\sum_{i=1}^{g} (\alpha_i^m + \alpha_i'^m) \right) \frac{x^m}{m} \right\}.$$

To facilitate the handling of the formulas for the number of rational points on $X_0(N)$ defined over $\mathbb{F}_{p^m}$ we now introduce some modification to the operators T_{p^m}: let

$$U(1) = 2Id$$

$$U(p) = T_p$$

$$U(p^m) = T_{p^m} - pT_{p^{m-2}}, \qquad m \ge 2.$$

Observe that for each m, the trace of $U(p^m)$ is given by

$$\operatorname{Tr} U(p^m) = \operatorname{tr} T_{p^m} - p \operatorname{Tr} T_{p^{m-2}}$$

$$= \sum_{i=1}^{g} \{\alpha_i^m + \alpha_i'^m\};$$

from this expression we readily obtain

$$H_p(x) = \exp \left\{ - \sum_{m=1}^{\infty} \operatorname{Tr} U(p^m) \frac{x^m}{m} \right\}.$$

The following is then a restatement of the congruence relation.

Corollary 5.10.1 *With notations as above, and p a prime which does not divide N and m a positive integer, we have*

$$\operatorname{card} X_0(N)(\mathbb{F}_{p^m}) = p^m + 1 - \operatorname{Tr} U(p^m).$$

5.6.8 The Eichler–Selberg trace formula

In this section we shall give an explicit formula for $\operatorname{Tr} U(p^f)$. This will be based on the following special case of Hijikata's version of the trace formula for the Hecke operators acting on the space of cusp forms of weight 2 for the group $\Gamma_0(N)$. Let $T_n = \prod_{p^a \| n} T_{p^a}$.

Theorem 5.11 *Let n be a positive integer relatively prime to N. We then have*

$$\operatorname{Tr} T_n = \sigma(n) + \delta(\sqrt{n})\mu_N - \sum_s a(s) \sum_f b(s,f)c(s,f),$$

where $\delta(\sqrt{n}) = 1$ if n is a square and 0 otherwise, μ_N is the index $[\Gamma : \Gamma_0(N)] =$

$N \prod_{l|N}(1 + 1/l)$ *and the meaning of the other symbols is given as follows. Let s run through all the integers such that*

$$d = s^2 - 4n$$

is not a positive non-square integer; hence for some positive integer t it has one of the following forms, designated respectively by (p), (h), (e), (e'):

$$d = \begin{cases} 0 & (p) \\ t^2 & (h) \\ t^2 m < 0 \text{ and } m \equiv 1 \ (\text{mod } 4) & (e) \\ 4t^2 m < 0 \text{ and } m \equiv 2, 3 \ (\text{mod } 4) & (e'). \end{cases}$$

For each of these we have

$$a(s) \begin{cases} |s|/8 & (p) \\ \frac{1}{2}\min\left(\left|1 + \frac{s}{t}\right|, \left|1 - \frac{s}{t}\right|\right) & (h) \\ \frac{1}{2} & (e, e'). \end{cases}$$

For each fixed s corresponding to its type, let f run as follows

$$f = \begin{cases} 1 & (p) \\ \text{all positive divisors of } t & (h, e, e') \end{cases}$$

and let

$$b(s, f) = \begin{cases} 1 & (p) \\ \frac{1}{2}\varphi(t/f) & (h) \\ h(d)/w(d) & (e, e') \end{cases}$$

where φ is Euler's function, $h(d)$ denotes the class number of primitive ideals of the order $\mathcal{O}_d$ of $\mathbb{Q}(\sqrt{d})$ with discriminant d and w is the order of the group of units $\mathcal{O}_d^\times$. For fixed s and f and a prime $l|N$, let $A_l(s, f)$ (resp. $B_l(s, f)$) be a complete system of representatives modulo $l^{\text{ord}_l(N)+\text{ord}_l(f)}$ for the set of integers $x \in \mathbb{Z}$ satisfying

$$x^2 - sx + n \equiv 0 \bmod l^{\text{ord}_l(N)+2\,\text{ord}_l(f)},$$

$$2x \equiv s \bmod l^{\text{ord}_l(f)},$$

(respectively the subset satisfying

$$x^2 - sx + n \equiv 0 \bmod l^{\text{ord}_l(N)+2\,\text{ord}_l(f)+1}).$$

Put

$$
c_l(s,f) = \begin{cases} \text{card } A_l(s,f) & \text{if } \dfrac{s^2 - 4n}{f^2} \not\equiv 0 \bmod l \\[3mm] \text{card } A_l(s,f) + \text{card } B_l(s,f) & \text{if } \dfrac{s^2 - 4n}{f^2} \equiv 0 \bmod l, \end{cases}
$$

and let

$$
C(s,f) = \prod_{l|N} c_l(s,f).
$$

Remark The most direct proof of the above theorem is that given by Eichler using Green's functions on the Riemann surface $X_0(N)$. Observe that the term $(\mu_0(N))/12$ is also the leading term in the expression for the genus of $X_0(N)$ for N large.

5.6.9 Proof of the theorem of TsFasman-Vladut and Zink

In the Eichler–Selberg trace formula we take N as a prime number and $n = p^2$. As indicated in the previous section, the congruence relation provides a count for the number of points on $X_0(N)$ rational over $\mathbb{F}_{p^2}$:

$$
\text{card } X_0(N)(\mathbb{F}_{p^2}) = p^2 + 1 - \text{Tr } U(p^2)
$$
$$
= p^2 + 1 + pg_0(N) - \text{Tr } T_{p^2}.
$$

From the formula for the trace of the Hecke operator we obtain

$$
\text{Tr } T_{p^2} = -\sum_{s,f} \frac{h(D)}{W(D)}\left(1 + \left(\frac{D}{N}\right)\right) + p^2 + 1 + g_0(N),
$$

where s runs over the integers satisfying $-2p < s < 2p$, f over the positive integers with f^2 a divisor of $s^2 - 4p^2$ so that

$$
D = \frac{s^2 - 4p^2}{f^2} \equiv 0 \qquad \text{or} \qquad 1 \bmod 4
$$

and (D/N) is the Legendre symbol. Substituting this into the above formula we obtain

$$
\text{card } X_0(N)(\mathbb{F}_{p^2}) = \sum_{s,f} \frac{h(D)}{w(D)}\left(1 + \left(\frac{D}{N}\right)\right) + g_0(N)(p - 1).
$$

As N becomes large, the sum $\sum_{s,f}$ remains bounded and the dominant term is $g_0(N)(p - 1)$. Thus we have the following theorem

Theorem 5.12 (TsFasman, Vladut, Zink) *For $q = p^2$ we have*

$$\lim_{N \to \infty} \frac{g_0(N)}{\operatorname{card} X_0(N)(\mathbb{F}_q)} = \frac{1}{\sqrt{q}-1}.$$

Remark The work of Ihara, Vladut and Drienfeld ([40], [18]), which uses more powerful techniques from algebraic geometry especially the theory of Shimura varieties and Drienfeld modules, leads to a stronger result. Let $q = p^{2m}$ be fixed, let $N(C)$ be the number of points on a curve C which are rational over $\mathbb{F}_q$ and let $g(C)$ be its genus. Then we have

$$\gamma(\mathbb{F}_q) = \lim \inf \frac{g(C)}{N(C)} = \frac{1}{\sqrt{q}-1},$$

where the lim inf is taken over all absolutely irreducible non-singular projective algebraic curves defined over $\mathbb{F}_q$. The proof in the case $q = p^4$ can also be obtained along the same lines as the argument given above for $q = p^2$, using Drienfeld's theory of elliptic modules together with his version of the Selberg trace formula in the function field case.

5.7 Examples of algebraic Goppa codes

In this section we describe several examples of codes constructed using the rational points on algebraic curves. Some of these have already been discussed by Goppa as part of his algebraico-geometric dictionary. The others have appeared in different disguises in papers by Serre, Driencourt and van Lint–Springer. Our presentation is sufficiently elementary for most of the material to be read independently of the rest of the chapter.

The algebraico-geometric dictionary of Goppa gives two approaches to the construction of codes which are dual to each other in a sense which is made precise in the theory of the residue theorem. We now describe the simpler approach which is based on the theory of linear systems. Let C be an algebraic curve of genus g and degree s embedded in projective $(r-1)$-space $\mathbb{P}^{r-1}$. Such a curve is the intersection of a number of hypersurfaces and can be realized as a linear series on a plane projective curve in $\mathbb{P}^2$ defined by a simple equation $F(x_0, x_1, x_2) = 0$, where x_0, x_1, x_2 are homogeneous coordinates. The linear series is defined by a sequence of homogeneous polynomials of the same degree $\varphi_i(x_0, x_1, x_2)$, $0 \le i \le r-1$:

$$\left\{ \sum_{i=0}^{r-1} a_i \varphi_i(x_0, x_1, x_2) = 0 : (a_0, \dots, a_{r-1}) \in P^{r-1} \right\}.$$

We suppose that the F and the φ_i have no components in common and that the linear series has no fixed points. The field of constants is assumed to be the finite field of q elements $\mathbb{F}_q$ and we suppose further that all the objects defined so far are all rational over $\mathbb{F}_q$. Letting y_i, $0 \le i \le r - 1$ be homogeneous coordinates for $\mathbb{P}^{r-1}$ and setting $\rho y_i = \varphi_i(x_0, x_1, x_2)$, we obtain a parametric representation of the curve C:

$$C \to \mathbb{P}^{r-1}, P \mapsto (\varphi_0(P), \ldots, \varphi_{r-1}(P)).$$

Let $Q_j = (\varphi_0(P_j), \ldots, \varphi_{r-1}(P_j))$, $1 \le j \le n$, be the points of C which are rational over $\mathbb{F}_q$. The parity check matrix of the algebraic Goppa code is defined by considering the coordinates of the point Q_j as the entries of the j-th column:

$$H = [\varphi_i(P_j)]_{\substack{0 \le i \le r-1 \\ 1 \le j \le n}}.$$

Remark For a description of the above construction as generalized Reed–Solomon codes see the paper by van Lint–Springer [92].

We are now ready to describe the various examples.

5.7.1 The Hamming $(7, 4)$ code

Shannon describes this code in his classic work [77] in the following terms. A basic block consists of seven symbols $x_1, x_2, x_3, x_4, x_5, x_6, x_7$. Of these x_3, x_5, x_6 and x_7 are the message symbols and are chosen arbitrarily by the source. The rest are the check symbols and are chosen so that the following relations are satisfied:

$$\alpha =: x_4 + x_5 + x_6 + x_7 \quad \text{even};$$

$$\beta =: x_2 + x_3 + x_6 + x_7 \quad \text{even};$$

$$\gamma =: x_1 + x_3 + x_5 + x_7 \quad \text{even}.$$

When a basic block is received, the α, β, γ are calculated and if even called zero, if odd called one. If at most one error has occured, then the binary number $\alpha\beta\gamma$ gives the location of the error. For instance, if the message symbols are $x_3 = 0$, $x_5 = 1$, $x_6 = 1$, $x_7 = 0$, then the check symbols are $x_4 = 0$, $x_2 = 1$, $x_1 = 1$ and the encoded message is 1100110. If the fifth symbol gets changed so that the received message is 1100010, then

$$\alpha =: 0 + 0 + 1 + 0 \equiv 1$$

$$\beta =: 1 + 0 + 1 + 0 \equiv 0$$

$$\gamma =: 1 + 0 + 0 + 0 \equiv 1$$

and the binary number 101 represents the fifth location as it should.

Over the binary field $\mathbb{F}_2$, the parity check matrix of this code is obtained by writing the binary expansions of the numbers 1, 2, 3, 4, 5, 6, 7 as the columns of a 3-by-7 matrix:

$$H = \begin{pmatrix} 0 & 0 & 0 & 1 & 1 & 1 & 1 \\ 0 & 1 & 1 & 0 & 0 & 1 & 1 \\ 1 & 0 & 1 & 0 & 1 & 0 & 1 \end{pmatrix}.$$

If the symbols x_i, $i = 1, \ldots, 7$ are interpreted as the coordinates of a vector in $\mathbb{F}_2^7$, then a code word, $x = (x_1, x_2, x_3, x_4, x_5, x_6, x_7)$ is in the null space of the matrix H, i.e. $H \cdot x = 0$. This is equivalent to the system

$$x_4 + x_5 + x_6 + x_7 = 0;$$

$$x_2 + x_3 + x_6 + x_7 = 0;$$

$$x_1 + x_3 + x_5 + x_7 = 0.$$

The geometric nature of this code comes about by observing that the projective plane $\mathbb{P}^2$ has precisely seven points rational over $\mathbb{F}_2$. The coordinates of these points coincide with the entries in the parity check matrix H. This is also the maximum number of rational points which a plane projective algebraic curve can have over $\mathbb{F}_2$. Serre has shown [74] that over $\mathbb{F}_2$ the maximum number of points which an algebraic curve of genus 3 can have is 7 and has given the following projective model:

$$C: x_0^3 x_1 + x_1^3 x_2 + x_2^3 x_0 + x_0^2 x_1^2 + x_1^2 x_2^2 + x_0^2 x_2^2 + x_0^2 x_1 x_2 + x_0 x_1^2 x_2.$$

In Goppa's scheme we take the linear system defined by the three functions $\varphi_0(x_0, x_1, x_2) = x_0$, $\varphi_1(x_0, x_1, x_2) = x_1$, $\varphi_2(x_0, x_1, x_2) = x_2$. If P_i, $i = 1, \ldots, 7$ runs over the seven rational points of C over $\mathbb{F}_2$ we obtain the parity check matrix as

$$H = (\varphi_i(P_j))_{\substack{i = 0, 1, 2 \\ j = 1, \ldots 7}}.$$

Remarks

1. The deeper aspects of this algebraic description are undoubtedly connected with the dual description in terms of differentials on the curve C.

2. There is an interesting description of the Hamming $(7, 4)$ code which uses the Boolean algebra generated by three sets.

5.7.2 BCH codes

In $\mathbb{P}^2$ we consider the locus C defined by the equation $x_2 = 0$. On C there is a linear system defined by the curves of degree $d - 1$, i.e. by the span of the functions:

$$\varphi_0 = x_1^{d-1}, \qquad \varphi_1 = x_1^{d-2}x_0, \ldots, \varphi_{d-1} = x_0^{d-1}.$$

This provides a map

$$\varphi: C \to \mathbb{P}^{d-1}.$$

The points on C rational over $\mathbb{F}_q$ are $(\alpha_j, 1, 0)$, $\alpha_j \in \mathbb{F}_q$, $1 \leq j \leq n$, and $(1, 0, 0)$. The parity check matrix corresponding to this linear system is

$$\begin{pmatrix} 1 & 1 & \cdots & 1 \\ \alpha_1 & \alpha_2 & \cdots & \alpha_n \\ \vdots & \vdots & \ddots & \vdots \\ \alpha_1^{d-1} & \alpha_2^{d-1} & \cdots & \alpha_n^{d-1} \end{pmatrix}.$$

In $\mathbb{P}^{d-1}$, every line intercepts $\varphi(C)$ in $d-1$ points and therefore any d points are not collinear and the weight of the code is $d + 1$.

Remark Information about the code parameters can be obtained from the internal geometry of the curve $\varphi(C)$ in $\mathbb{P}^{d-1}$. This is to some extent elucidated in the approach to codes via finite geometries.

5.7.3 The Fermat cubic (Hermite form)

Over the finite field $\mathbb{F}_4 = \{1, \alpha, \beta, 0\}$ the Fermat cubic

$$C: X^3 + Y^3 + Z^3 = 0$$

has the structure of a Hermitian form

$$X \cdot \bar{X} + Y \cdot \bar{Y} + Z \cdot \bar{Z} = 0,$$

where $\sigma: X \mapsto \bar{X}$ is the automorphism of order two of $\mathbb{F}_4$, i.e. the generator of $\text{Gal}(\mathbb{F}_4/\mathbb{F}_2)$ given by $\sigma(X) = X^2$. This curve has a rich arithmetic and geometric structure which we recall briefly below.

As a moduli variety, C is a model for the modular curve $X_0(27)$. Since C is non-singular, its genus is clearly $g = 1$. The number of points on C over $\mathbb{F}_4$ is 9; these are in fact the 3-division points on C, i.e. the kernel of the isogeny 'multiplication by 3' [3]: $C \mapsto C$. These points have a very special arrangement. It is well known (Serre [75], p. 65) that the space of cusp forms of weight 2 for $\Gamma_0(27)$ is of dimension 1 and its normalized generator is given by

$$d\omega = q \prod_{n=1}^{\infty} (1 - q^{3n})^2 (1 - q^{9n})^2$$

$$= \sum_{n=1}^{\infty} a(n) q^n.$$

We have already indicated in Section 5.4.2 that the Fermat curve has a Weierstrass model of the form $Y^2 - Y = X^3 - 7$. This curve is also isogenous to the curve $Y^2 + Y = X^3 - 3$ given in Serre [75]. The points on C rational over $\mathbb{F}_4$ are:

$$
\begin{array}{c}
 \quad P_1 \; P_2 \; P_3 \; P_4 \; P_5 \; P_6 \; P_7 \; P_8 \; P_9 \\
\begin{array}{c} x \\ y \\ z \end{array}
\left(
\begin{array}{ccccccccc}
1 & 0 & \alpha & 1 & 0 & \beta & 0 & \alpha & 1 \\
0 & 1 & 0 & 1 & \alpha & 0 & \beta & 1 & \alpha \\
1 & 1 & 1 & 0 & 1 & 1 & 1 & 0 & 0
\end{array}
\right).
\end{array}
$$

These points are also the points of inflection of the curve

$$C: F = X^3 + Y^3 + Z^3 = 0,$$

that is to say the points of intersection of C with the curve

$$C^*: F^* = \det\left(\frac{\partial^2 F}{\partial x_i \partial x_j}\right) = xyz = 0.$$

We note that the formal derivatives of F are calculated in the sense of Hasse so that $\partial^2 F/\partial^2 x = x$, etc. Thus the Hessian of the Fermat curve consists of the three lines at infinity on the projective plane.

To define the code associated with this eliptic curve, we consider the linear system of all conics in $\mathbb{P}^2$:

$$\lambda_0 x_0^2 + \lambda_1 x_1^2 + \lambda_2 x_2^2 + \lambda_3 x_0 x_1 + \lambda_4 x_1 x_2 + \lambda_5 x_0 x_2 = 0.$$

These form a projective space of dimension 5. The resulting embedding is

$$\varphi: C \hookrightarrow \mathbb{P}^5, \qquad P \mapsto (\varphi_0(P), \ldots, \varphi_5(P))$$

where

$$\varphi_0 = x_0^2, \; \varphi_1 = x_1^2, \; \varphi_2 = x_2^2, \; \varphi_3 = x_0 x_1, \; \varphi_4 = x_1 x_2, \; \varphi_5 = x_0 x_2.$$

The parity check matrix $(\varphi_i(P_j))$, $i = 0, \ldots, 5, j = 1, \ldots, 9$ for this code is

$$
\begin{pmatrix}
1 & 0 & \alpha & 1 & 0 & \alpha & 0 & \beta & 1 \\
0 & 1 & 0 & 1 & \beta & 0 & \alpha & 1 & \beta \\
1 & 1 & 1 & 0 & 1 & 1 & 1 & 0 & 0 \\
0 & 0 & 0 & 1 & 0 & 0 & 0 & \alpha & \alpha \\
0 & 1 & 0 & 0 & \alpha & 0 & \beta & 0 & 0 \\
1 & 0 & \alpha & 0 & 0 & \beta & 0 & 0 & 0
\end{pmatrix}.
$$

The dehomogenization of the linear system above using the conic $\psi = x_0 x_1 + x_1 x_2 + x_2 x_0$, which takes nonzero values $(1, 1, \alpha, 1, \alpha, \beta, \beta, \alpha, \alpha)$ at the points P_j, leads to the canonical form of the parity check matrix

$$H = \begin{pmatrix} 1 & 0 & 0 & 0 & 0 & 0 & \alpha & \alpha & 0 \\ 0 & 1 & 0 & 0 & 0 & 0 & \alpha & \beta & 1 \\ 0 & 0 & 1 & 0 & 0 & 0 & \beta & 0 & \alpha \\ 0 & 0 & 0 & 1 & 0 & 0 & 0 & 1 & 1 \\ 0 & 0 & 0 & 0 & 1 & 0 & \beta & \beta & 1 \\ 0 & 0 & 0 & 0 & 0 & 1 & 1 & \alpha & \alpha \end{pmatrix}.$$

from which we obtain the generator matrix

$$G = \begin{pmatrix} \alpha & \alpha & \beta & 0 & \beta & 1 & 1 & 0 & 0 \\ \alpha & \beta & 0 & 1 & \beta & \alpha & 0 & 1 & 0 \\ 0 & 1 & \alpha & 1 & 1 & \alpha & 0 & 0 & 1 \end{pmatrix}.$$

Remark The symmetry group of this code is the so-called Hessian group which describes the relative position of the nine 3-division points on the Fermat cubic.

5.7.4 Elliptic codes (according to Driencourt–Michon)

A detailed analysis of a very interesting class of Goppa codes associated with curves of genus 1 over the binary field has been made by Driencourt and Michon [20] We want to introduce here certain aspects of the construction from an elementary point of view.

We consider only those elliptic curves defined by equations of the type

$$E: y^2 + y = u(x),$$

where $u(x)$ is a polynomial in $\mathbb{F}_2[x]$. The number $N_2(E)$ of $\mathbb{F}_2$-rational points on E is 1, 2 or 5. In general over the field $\mathbb{F}_q$, $q = 2^m$, the number of $\mathbb{F}_q$-rational points is a quantity expressible in terms of $N_2(E)$. Here we need to use the fact that this number is odd and that all points except the point at infinity $P_\infty = (0, 1, 0)$ are paired as follows

$$P = (\alpha, \beta, 1) \leftrightarrow \bar{P} = (\alpha, \beta + 1, 1).$$

Let A be the subset of $\mathbb{F}_q$ consisting of the distinct non-zero x-coordinates of $\mathbb{F}_q$-rational points of E. Then the complete set of $\mathbb{F}_q$-rational points on E is

$$P_\infty, P_\alpha = (\alpha, \beta, 1), \qquad \bar{P}_\alpha = (\alpha, \beta + 1, 1), \qquad \alpha \in A.$$

We let $a = \text{card } A$. For a positive integer d, we define two divisors of E:

$$G = dP_\infty, \qquad D = \sum_{\alpha \in A} (P_\alpha + \bar{P}_\alpha).$$

We suppose that $2a - d \geq 1$. It is clear that the support of both divisors are disjoint and that deg $D = 2a$. Put $\delta = [d/2], \delta^* = [(d-3)/2]$ and define a function on E by

$$f(x, z) = \prod_{\alpha \in A} (x + \alpha z).$$

$$\omega_j, 0 \leq j \leq a - \delta - 2; \qquad \omega_j^*, 0 \leq j \leq a - \delta^* - 2$$

where

$$\omega_j = x^j z^{a-1-j} y \frac{dx}{f}, \qquad \omega_j^* = x^j z^{a-j} \frac{dx}{f}.$$

Using the fact that $T = x + \alpha$ can serve as a local uniformizing parameter at $P = (\alpha, y(\alpha), 1)$, one calculates readily that the residues of the corresponding differentials are given by

$$\text{Res}_P\left(x^l \frac{dx}{f}\right) = \frac{\alpha^l}{f'(\alpha)}, \qquad \text{Res}_P\left(x^l y \frac{dx}{f}\right) = \frac{\alpha^l y(\alpha)}{f'(\alpha)},$$

where

$$f'(\alpha) = \prod_{\substack{\alpha' \in A \\ \alpha' \neq \alpha}} (\alpha + \alpha').$$

With $n = 2a$, we define the residue map

$$\text{Res}: \Omega(G - D) \to \mathbb{F}_q^n,$$

where

$$\omega \mapsto \text{Res}(\omega) = (\text{Res}_{P_1}\omega, \ldots, \text{Res}_{P_a}\omega, \text{Res}_{\bar{P}_1}\omega, \ldots, \text{Res}_{\bar{P}_a}\omega).$$

The resulting Goppa code is the linear subspace of $\mathbb{F}_q^n$ generated by the vectors $\text{Res}(\omega)$, as ω varies over $\Omega(G - D)$. The parameters of this code are $n = 2a, k = 23 - d$ and *weight* $\geq d$.

Remark The elliptic Goppa codes over binary and ternary fields are related to very interesting number theoretic sums, such as that of Kloostermann. The interested reader would do well to consult the original papers of Driencourt–Michon [20] Lachaud–Wolfmann [48], Livne [51].

5.7.5 The Klein quartic

The modular curve $X(7) = H/\Gamma(7) \cup Q/\Gamma(7)$ associated with the principal congruence subgroup of level 7 has genus 3. It is non-hyperelliptic and a normal embedding into the projective plane $\mathbb{P}^2$ can be achieved by the mapping

$$X(7)_C \hookrightarrow \mathbb{P}^2(C), \ P \mapsto (\omega_0(P), \omega_1(P), \omega_2(P)),$$

where ω_i, $i = 1, 2, 3$, are a basis of the 3-dimensional space of holomorphic differentials on $X(7)$. Using the fact that the space of cusp forms of weight two for the group $\Gamma(7)$ has a basis whose Fourier coefficients lie in the cyclotomic field $Q(\zeta_7)$ it is possible to show that $X(7)$ has a plane model defined over the same field. In fact it can be shown that there is a plane model defined over the ring $\mathbb{Z}$ of rational integers and given by:

$$K: X^3Y + Y^3Z + Z^3X = 0.$$

This curve was extensively studied by Klein because of its connection with automorphic forms, complex multiplication and its large group of automorphisms. An excellent exposition from the modern arithmetic point of view is given in the survey article by Mazur [60]. We recollect here some of its more important properties. The plane model K has good reduction modulo all primes other than 7, and at 7 it is possible to have the ramification absorbed by a sufficiently ramified number field; that is, there are number fields over which K has good reduction everywhere. The Jacobian of K is a product of three elliptic curves all isogenous to the elliptic curve with complex multiplication over the field $Q(\sqrt{-7})$. This last fact implies the existence of a formula for counting the number of rational points on K over the field $\mathbb{F}_p$ which depends on how the prime p splits in the ring $\mathbb{Z}[\sqrt{-7}]$. The automorphism group is isomorphic to $PSL_2(\mathbb{F}_7)$, the simple group of order 168, which according to Hurwitz theorem is the largest possible order for the automorphism group of an algebraic curve of genus 3 in characteristic zero. Since $\mathbb{F}_8$ has a primitive seventh root of 1, and since all the automorphisms of K are rational over $Q(\zeta_7)$, we also get the same automorphisms in characteristic 2, rational over $\mathbb{F}_8$. It is also interesting that the Klein quartic achieves the maximum number of rational points over $\mathbb{F}_8$ that is allowed by Serre's bound:

$$|\text{card } K(\mathbb{F}_q) - (q + 1)| \leq g[2\sqrt{q}],$$

i.e. card $K(\mathbb{F}_8) = 24$. To describe explicitly these rational points we consider the following automorphisms:

$$\sigma(x, y, z) = (\zeta x, \zeta^4 y, \zeta^2 z), \eta(x, y, z) = (z, x, y),$$

where ζ is a generator of the cyclic group $\mathbb{F}_8^\times$. These generate a group $G = \langle \sigma, \eta \rangle$ of order 21. To list the rational points we start with:

$$Q_0 = (1, 0, 0), \ Q_1 = (0, 1, 0), \ Q_2 = (0, 0, 1).$$

The remaining are obtained from the point $P_{00} = (1, \zeta^2, \zeta^2 + \zeta^2)$ by the action:

$$P_{ij} = \eta^i \sigma^j P_{00}, \ i = 0, 1, 2; j = 0, \ldots, 6,$$

where we use the fact that $\zeta^3 + \zeta + 1 = 0$. This set of points splits under the action of the group G into two orbits, one formed by the Q's and the other formed by the P's.

There are many Goppa codes that can be constructed from the Klein quartic. We describe below a simple concrete example which exhibits most of the peculiarities of the general situation. We start with an integer m satisfying the condition $2 \leq m \leq 6$ and consider the polar divisor

$$D = m(Q_0 + Q_1 + Q_2).$$

As in the Riemann–Roch theorem, let $L(D)$ be the vector space of functions φ such that $\varphi \equiv 0$ or $\mathrm{div}(\varphi) \geq -D$, that is the poles of φ are supported on the set $\{Q_0, Q_1, Q_2\}$ and are of multiplicity at most m. We now consider the mapping:

$$K \rightarrow \mathbb{P}^{l-1}, \ P \mapsto (\varphi(P))_{\varphi \in L(D) - \{0\}},$$

where $l = \dim_{F_8} L(D) = 3m - 2$. The parity check matrix of the code is

$$H = (\varphi(P)),$$

where the columns are indexed by the points $P \in K(\mathbb{F}_8) - \{Q_0, Q_1, Q_2\}$ and the rows are indexed by the functions φ in $L(D) - \{0\}$. The linear code generated is a subspace of $\mathbb{F}_8^{21}$ of dimension $3m - 2$. In the particular case $m = 4$, $\dim L(D) = 10$ and a basis for $L(D)$ is given by the 10 rational functions:

$$\frac{x^3}{xyz}, \ \frac{x^2 y}{xyz}, \ \frac{x^2 z}{xyz}, \ \frac{xy^2}{xyz}, \ \frac{xyz}{xyz}, \ \frac{xz^2}{xyz}, \ \frac{y^3}{xyz}, \ \frac{y^2 z}{xyz}, \ \frac{yz^2}{xyz}, \ \frac{z^3}{xyz}.$$

As to the computation of the code parameters, we observe that each function φ can have at most $\deg(D) = 3m$ zeros. Therefore in a code vector $(\varphi(P))_{P \in K(\mathbb{F}_8) - \{0\}}$ there can be at most $3m$ zero components. This implies that the weight of the code is at least $21 - 3m$ and in the case under consideration the weight is then at least 9. This is then the code $(21, 10, 9)$. We leave the discussion of many other interesting aspects of codes on the Klein quartic to the reader. Possibly the most notable omission is an analysis of the action of the automorphism group on the codes and its effect on the weight distribution.

Remarks 1 Manin and Vladut [55] have constructed Goppa codes using Drienfeld modules. These codes have very good parameters and their space and time complexity have been estimated.

2. In a very interesting article [93] Vasquez has discussed a family of

codes arising from the curves $y^a = x^q + x$. His construction is fairly elementary and some of the codes have a large symmetry group.

3. In his Yale thesis [65] Sid Porter has developed symbolic manipulation techniques to handle the decoding problem for Goppa codes. The essential tool in his method is a diophantine approximation theorem that works over function fields of curves of positive genus. In her CUNY thesis, Despina Polemis has made a study of the effective constructibility of Goppa codes using Noether's theorem of the residue.

Exercises

1. The elliptic curve $y^2 + y = x^3$ in characteristic 2 has the interesting property that its group of automorphisms coincides with the unit group of the integral Hurwitz quaternions. Such a group is isomorphic to $SL_2(\mathbb{F}_3)$ and is the group of smallest order which admits an irreducible representation of degree 2 which is non-monomial. Study the infinite family of elliptic Goppa codes over the tower of fields $\mathbb{F}_{2^m}$, $m = 1, 2, \ldots$. Determine their weight distribution.

2. The Fermat elliptic curve $x^3 + y^3 + z^3 = 0$ is birationally equivalent to C: $y^2 + y = x^3 + 1$ over $\mathbb{F}_2$. Verify that the rational points of C over $\mathbb{F}_4$ are the point at infinity $P_\infty(0, 1, 0)$ and the eight points

$$P_0 = (0, \alpha, 1), \qquad P_\alpha = (\alpha, 1, 1), \qquad P_\beta = (\beta, 1, 1), \qquad P_1 = (1, 1, 1)$$

$$\bar{P}_0 = (0, \beta, 1), \qquad \bar{P}_\alpha = (\alpha, 0, 1), \qquad \bar{P}_\beta = (\beta, 0, 1), \qquad \bar{P}_1 = (1, 0, 1).$$

For each allowable integer d, construct the residue map

$$\text{Res: } \Omega(G - D) \rightarrow \mathbb{F}_4^8.$$

Observe that the resulting Goppa codes are different from the Fermat elliptic code constructed above.

Appendix
Simplification of the singularities of algebraic curves

The classical study of diophantine equations like that of Fermat has come to depend on the properties of algebraic curves over fields of characteristic p; in analogy with the algebraic geometry of curves over the field of complex number, where the theory of Riemann surfaces is very suggestive, it is customary to assume that the field of constants is algebraically closed. For example the original proof of the Riemann hypothesis for function fields was carried out entirely within the geometric context of algebraic surfaces. When dealing with concrete examples one desires a more direct approach which depends more on the arithmetic of finite fields and the algebra of polynomials. For instance the general theory implies that there is a simple formula for counting the number of solutions on the Klein curve

$$XY^3 + YZ^3 + ZX^3 = 0$$

which are rational over the finite field $\mathbb{F}_p$; this formula can be made explicit in terms of the behavior of the prime ideal (p) in the ring of integers of the complex quadratic field of discriminant -7. Besides finding the points on a curve which are rational over a finite field, one is also interested in solving the Riemann–Roch theorem arithmetically, e.g. when the curve is defined by a given polynomial, say in two variables, one wants to effectively construct functions and differentials with prescribed poles and zeros. In many situations it is impractical to suppose at the beginning that the given curve is non-singular or even irreducible. Aside from their intrinsic number theoretical interest, the answers to some of the questions raised above have become of some significance in the theory of error correcting codes owing to the discovery by Goppa that interesting families of linear codes can be constructed inside linear systems associated with algebraic curves. In his paper [31], Goppa has elaborated a dictionary based on the classical algebraic geometry of curves which leads to the constructions and calculations necessary for applications to error correcting codes. These suggestions of Goppa, viewed from the optic of modern algebraic geometry, have to be reconciled with the possibility that the field of constants is not algebraically closed; the notion of a point has to be clarified, especially when the curve is allowed to have singularities. It is the experience of the author that almost all the facts about algebraic curves which are necessary for developing the ideas of Goppa on a solid and constructive basis have been available for over a hundred years, some of them in fact going back to Riemann and Noether;

unfortunately to find these ideas one often needs to consult more than one classical treatment. It is our intention to present in this appendix a coherent introduction to the theory of algebraic curves over finite fields. As we do not intend to survey the entire field, we have taken as the central theme a presentation of the theorem which ascertains that a plane model can be transformed by a finite number of quadratic transformations into a model whose singularities are at worst ordinary in the sense that all the tangents at a multiple point are distinct. This is the main step in several important results: the calculation of the genus, the existence of a desingularization, the formula for the intersection index of two curves, the construction of a basis for linear systems using Noether's theorem of the residue, etc.

The contents of the notes are as follows. In Sections A.1 and A.2 we introduce the elementary notions about projective coordinates and prove a few elementary lemmas about the dimension of certain families of curves. In Section A.3 we recall the definition of a dual curve and recall some basic results associated with the so-called Plücker formulas. In Section A.4 we begin the study of quadratic transformations and set the stage for the proof of the main result about the reduction of singularities given in Sections A.4.1–A.4.4.

Our presentation, which was originally motivated by the desire to understand the suggestions of Goppa, has been greatly influenced by that found in the two classics on algebraic curves by Walker [94] and Fulton [26].

A.1 Homogeneous coordinates in the plane

In affine two-space $\mathbb{A}^2$ over a field k, we consider a triangle formed by three linear forms $L_i = a_i x + b_i y + c_i$, $i = 0, 1, 2$, in $k[x, y]$ with a non-zero determinant $D = (a_i b_i c_i)$, $i = 0, 1, 2$, over a field k of characteristic zero. With a fixed choice of square roots, the formulas

$$p_i = (a_i x + b_i y + c_i)/(a_i^2 + b_i^2)^{1/2}, \qquad i = 0, 1, 2$$

can be interpreted as representing 'the distance' from the point $(x, y) \in \mathbb{A}^2$ to the lines L_0, L_1, L_2. This terminology suggests the introduction of homogeneous coordinates for the point (x, y) given by the three numbers x_0, x_1, x_2 such that

$$x_0 = k_0 p_0, \qquad x_1 = k_1 p_1, \qquad x_2 = k_2 p_2,$$

where k_0, k_1, k_2 are fixed constants, the same for all points and p_i is a factor of proportionality. Without loss of generality we suppose in the following that $k_0 = k_1 = k_2 = 1$; we thus write

$$x_i = a_i x + b_i y + C_i, \qquad i = 0, 1, 2; \tag{A.1}$$

solving for x and y we obtain

$$Dx = \sum_{i=0}^{2} A_i x_i, \qquad Dy = \sum_{i=0}^{2} B_i x_i, \qquad D = \sum_{i=0}^{2} C_i x_i,$$

or equivalently

$$x = \left(\sum_{i=0}^{2} B_i x_i \right) \bigg/ \left(\sum_{i=0}^{2} C_i x_i \right), \qquad Y = \left(\sum_{i=0}^{2} B_i x_i \right) \bigg/ \left(\sum_{i=0}^{2} C_i x_i \right).$$

These equations give a concrete way of transforming an affine model $f(x, y)$ of a curve in $\mathbb{A}^2$ into a projective model $F(x_0, x_1, x_2)$ in $\mathbb{P}^2$, where we have the freedom to place the line at infinity in a desirable position.

Example Consider the curve. $C: f(x, y) = y^2 - y - x^3 + x^2$ over the field $\mathbb{F}_{11}$. In $\mathbb{A}^2$ we consider the triangle defined by the linear forms

$$L_0: y - 2x - 5 = 5, \qquad L_1: y + x - 5 = 0, \qquad L_3: y + 6x - 3 = 0.$$

L_0 and L_2 pass through the point $Q = (8, 6)$, where the curve looks like a node with two distinct tangents.

Solving for X and Y we obtain:

$$x = (-2x_0 + 2x_1)/(5x_0 - 8x_1 + 3x_2),$$

$$y = (5x_0 - 3x_1 + 4x_2)/(5x_0 - 8x_1 + 3x_2);$$

substituting these in the affine model $f(x, y)$ and simplifying we obtain the equation for the projective model $F(x_0, x_1, x_2)$.

If we choose a different set of lines $L_i' = a_i'x + b_i'y + c_i'$, $i = 0, 1, 2$, with non-zero determinant $D' = (a_i'b_i'c_i')$ $i = 0, 1, 2$ we are led to a new set of homogeneous coordinates Y_0, Y_1, Y_2 given by the formulas

$$Y_i = a_ix + b_iy + c_i', \qquad i = 0, 1, 2.$$

Solving these for x and y and substituting into the equations (A.1) we obtain the relations

$$X_i = C_{i0}Y_0 + C_{i1}Y_1 + C_{i2}Y_2, \qquad i = 0, 1, 2, \qquad \det(C_{ij}) \neq 0.$$

Hence a change of triangle in $\mathbb{A}^3$ simply corresponds to a linear projective change of coordinates in $\mathbb{P}^2$.

Remark As we shall see later on, Euler's theorem on homogeneous polynomials together with the fact that two distinct homogeneous coordinate systems in $\mathbb{P}^2$ are related by a projective linear transformation implies that the structure of a singularity of a plane projective curve is independent of the coordinate system chosen.

A.2 Basic lemmas

Definition A.1 Let $F(x_0, x_1, x_2)$ be a homogeneous polynomial of degree d in x_0, x_1, x_2; let $C : F(x_0, x_1, x_2) = 0$ be the associated curve.

To say that $P = (1, 0, 0)$ *is a point on* C *of multiplicity* m *means that when expressing* F *as a polynomial in powers of* x_0 *with coefficients which are homogeneous polynomials in* X_1, X_2, *we have*

$$F = F_m(X_1, X_2)X_0^{d-m} + F_{m+1}(X_1, X_2)X_0^{d-m-1} + \cdots F_d(X_1, X_2),$$

i.e., m *is the smallest degree of any of the coefficients* F_i.

As the above definition is clearly invariant under projective transformations, by translating an arbitrary point $Q = (a, b, c)$ to $P = (1, 0, 0)$, we arrive at a definition of what it means for Q to be a point on C of multiplicity m. This definition has the merit of being independent of the concept of derivative which works well mostly in characteristic 0. In addition it should be pointed out that if Q is a point with coordinates which are rational over the finite field $\mathbb{F}_q$ then the translation of Q to P does not necessitate an extension of the field of constants. On the other hand if we interpret the linear factors of the polynomial $F_m(x_1, x_2)$ as corresponding to the tangents at $P = (1, 0, 0)$ taken with appropriate multiplicities, it may become necessary to make a finite extension of $\mathbb{F}_q$, if one wants the tangents to be also rational over the field of constants.

The arguments below depend on dimension counts which we now proceed to explain. As we are dealing for the most part with homogeneous polynomials in x_0, x_1, x_2 which define algebraic curves in the projective plane $\mathbb{P}^2$ we first observe that the dimension of the vector space V spanned by the monomials $x_0^i x_1^j x_2^k$, with positive integers satisfying $i + j + k = d$ is $\frac{1}{2}d(d + 3) + 1$; hence the projective space $P(V)$ generated by the lines in V passing through the origin parametrizes the projective plane curves $C : F(x_0, x_1, x_0) = 0$ and has dimension $N = \frac{1}{2}d(d + 3)$. If $P = (a, b, c)$ is a point in $\mathbb{P}^2$, then the requirement that $F(P) = 0$ is equivalent to one linear condition and thus the dimension of the subspace of such polynomials is $N - 1$. In general, if it is required that P be a point of multiplicity m, then essentially $\frac{1}{2}m(m + 1)$ linear conditions have to be satisfied. In fact, if the point in question is $P = (1, 0, 0)$ and the representation of a polynomial $F(x_0, x_1, x_2)$ of degree d in powers of x_0 is given by

$$F = \sum_{j=0}^{d} F_j(x_1, x_2) x_0^{d-j},$$

then it is required that each polynomial $F_j(x_1, x_2)$ be identically zero for $j = 0, \ldots$, $m - 1$. Since each such polynomial has $j + 1$ free coefficients, the totality of conditions is $1 + 2 + \cdots + m = \frac{1}{2}m(m + 1)$. Thus the dimension of the linear space of projective plane curves of degree d having P as a point of multiplicity m is $\frac{1}{2}d(d + 3) - \frac{1}{2}m(m + 1)$.

Definition A.2 *Let $D = \sum_P m_P \cdot P$ be a positive cycle in $\mathbb{P}^2$. Let d be a positive integer. $V_{d,D}$ is linear subspace of $\mathbb{P}^n$ consisting of curves $C : F = 0$ of degree d such that the multiplicity of P as a point of C is $\geq m_P$ for all points P in the support of the divisor D.*

The argument given above gives immediately the results:

Theorem A.1 *With notations as above, we have*

$$\dim V_{d,D} \geq \frac{1}{2}d(d + 3) - \frac{1}{2}\sum_P m_P(m_P + 1).$$

A more careful counting argument actually gives the results:

Theorem A.2 *With notations as above if* $D \geq (\sum_P m_P) - 1$, *then*

$$\dim V_{d,D} = \frac{1}{2}d(d+3) - \frac{1}{2}\sum_P m_P(m_P + 1).$$

Remark For a proof of this result see Fulton [26].

Lemma A.1 *If* C: $F(x_0, x_1, x_2) = 0$ *is a projective plane curve of a degree* d *without multiple components and with* $P_1, \ldots, P_s$ *as points of respective multiplicities* $m_1, \ldots, m_s$, *then we have*

$$\sum_{i=1}^{s} m_i(m_i - 1) \leq d(d-1).$$

Proof. As the number of points in $\mathbb{P}^2$ rational over $\mathbb{F}_q$ is $q^2 + q + 1$, and since a projective plane curve C: $F(x_0, x_1, x_2) = 0$ of degree d can have at worst $dq^2/(q-1)$ rational points, it is possible, on the assumption that $q^2 + q + 1 \geq dq^2/(q-1) + 3$, to choose a projective coordinate system for $\mathbb{P}^2$ based on a triangle which does not have any of its vertices lying on C. If we write the polynomial

$$F = F_0(X_1, X_2)X_0^d + F_1(x_1, x_2)x_0^{d-1} + \cdots + F_{d-1}(x_1, x_2)x_0 + F_d(x_1, x_2),$$

where each F_i is homogeneous of degree i in x_1, x_2, then the polynomial

$$\tilde{F} = dF_0(x_1, x_2)x_0^{d-1} + (d-1)F_1(x_1, x_2)x_0^{d-2} + \cdots + F_{d-1}(x_1, x_2)$$

is homogeneous of degree $d - 1$ and not identically zero, since the point $P = (1, 0, 0)$ does not belong to C. Furthermore $\tilde{F}$ has P_i, $i = 1, \ldots, s$ as points of multiplicity at least $m_i - 1$.

The claim is then a consequence of the following weak form of Bezout's theorem, which can be obtained by an ellimination argument using resultants.

Proposition A.1 *Two projective plane curves* C, C' *of degrees* d, d' *and with no common components and with multiplicities* m_i, m_i' *at their points of intersection satisfy the relation*

$$\sum_{i=1}^{s} m_i m_i' \leq dd'.$$

Remarks 1. If the curve C: $F = 0$ is defined by a polynomial F which is the product of d distinct linear factors all passing through P, then the inequality in Lemma 2.5 cannot be improved.

2. It is important to observe that the assumption that C has no multiple components excludes the possibility that

$$F(x_0, x_1, x_2) = F^*(x_0, x_1, x_2)^{p^i}$$

for some power p^i. Otherwise the use of the formal derivative $\tilde{F}$ would be invalidated.

Definition A.3 *Let* $C: F(x_0, x_1, x_2) = 0$ *be a projective plane curve of degree d. The deficiency of C is defined by*

$$g^*(C) = \frac{1}{2}(d-1)(d-2) - \frac{1}{2}\sum_P m_P(m_P - 1),$$

where the sum $\sum_P$ runs over all the points P on C and m_P is its multiplicity.

Theorem A.3 *The deficiency of an irreducible projective plane curve is always non-negative, i.e.*

$$\frac{1}{2}(d-1)(d-2) \geq \frac{1}{2}\sum_P m_P(m_P - 1).$$

Proof. Since C has no multiple components, we have by Lemma A.1

$$\frac{1}{2}\sum_P m_P(m_P - 1) \leq \frac{1}{2}d(d-1) < \frac{1}{2}(d-1)(d+2),$$

hence there is a projective plane curve C' of degree $d - 1$ having at P a point of multiplicity $m_P - 1$ and passing through

$$\frac{1}{2}(d-1)(d+2) - \frac{1}{2}\sum_P m_P(m_P - 1)$$

simple points of C. Since C is irreducible and C' is of smaller degree, the two curves have no component in common and thus by Theorem A.1

$$d(d-1) \geq \sum_P m_P(m_P - 1) + \frac{1}{2}(d-1)(d+2) - \frac{1}{2}\sum_P m_P(m_P - 1),$$

hence the result.

A.3 Dual curves

Let $C: F(x_0, x_1, x_2) = 0$ be an irreducible projective plane curve in $\mathbb{P}^2$ of $\deg(F) \geq 2$ and with $K = \mathbb{F}_q(C)$ as its function field. The space of lines in $\mathbb{P}^2$ is in one-to-one correspondence with the points in $\mathbb{P}^2$: to a point (a_0, a_1, a_2) there is associated the line

$$L: a_0 x_0 + a_1 x_1 + a_2 x_2 = 0,$$

and conversely with L on associates the point (a_0, a_1, a_2). This suggests that each simple point P on the curve C one associates the point Q in $\mathbb{P}^2$ whose coordinates correspond to the tangent line to C at P. As we shall see below the closure of the set of all points Q in $\mathbb{P}^2$ is an irreducible projective plane curve with the same function field as C.

Since F is irreducible, it is not a power of another polynomial and hence we can choose a projective coordinate system so that the three formal derivatives

$$F_i = F_{x_i}, \qquad i = 0, 1, 2$$

are not simultaneously identically zero. Let U_i, $i = 0, 1, 2$, be the images of F_i, $i = 0$,

1, 2 in the homogeneous coordinate ring

$$\alpha: \mathbb{F}_q[X_0, X_1, X_2] \to \mathbb{F}_q[X_0, X_1, X_2]/(F)$$

$$X_i \to U_i \equiv F_i \bmod F.$$

Let I be the kernel of this mapping. It is a homogeneous prime ideal; in fact if f_1 and f_2 belong to I and are relatively prime, then $f_1 A + f_2 B = 1$ and both $f_1(F_0, F_1, F_2)$ and $f_2(F_0, F_1, F_2)$ are divisible by F which is impossible; also I is generated by an irreducible homogeneous polynomial $\tilde{F}$. The mapping

$$\alpha: \mathbb{F}_q[X_0, X_1, X_2]/(\hat{F}) \to \mathbb{F}_q[x_0, x_1, x_2]/(F)$$

induces at the level of the field of quotients a mapping

$$\hat{\alpha}: \mathbb{F}_q(\hat{C}) \to \mathbb{F}_q(C),$$

where $\mathbb{F}_q(C)$ is the function field of the curve $\hat{C}: \hat{F} = 0$. To verify that $\hat{C}$ is indeed a curve, we need to show that the rational functions F_1/F_0 and F_2/F_0 are not both constant. This is easily seen from the fact that the degrees of $F_1 - aF_0$ and $F_2 - bF_0$ with a and b constant is less than that of F and thus cannot be divisible by F. Also from Euler's theorem we have

$$F = X_0 F_0 + X_1 F_1 + X_2 F_2$$

$$= (X_0 + aX_1 + bX_2)F_0.$$

Since F is irreducible, the above equality implies $\deg F = 1$ and F_0 is a constant contrary to the assumption that F is not a line. This proves that the dual curve $\hat{C}: \hat{F} = 0$ is irreducible.

Definition A.4 *The class of an irreducible projective plane curve is the degree of its dual curve.*

Problem A.1 Given a projective plane curve $C: F = 0$, find an efficient algorithm to compute the homogeneous polynomial $\hat{F}$ which defines the dual curve $\hat{C}: \hat{F} = 0$.

A.3.1 Plucker formulas

Let $C: F(X_0, X_1, X_2) = 0$ be a plane projective curve with no multiple components; let $Q = (q_0, q_1, q_2)$ be a point in the complement of C in $\mathbb{P}^2$. Suppose Q lies on m lines that are tangent to C and that each of these lines is tangent to C at exactly one point which is neither a singularity nor an inflexion point; recall that this latter point is one where the tangent line has an order of contact at least 3. The degree of the dual curve $\hat{C}$ is clearly m, i.e. the class of C is m. To take this as a definition it must be shown that such points Q exist and that m is independent of Q. Let $P = (p_0, p_1, p_2)$ be a single point on C with coordinates rational over $\mathbb{F}_q$. Recall that a necessary and sufficient condition for Q to lie on the tangent to C at P is that

$$\sum_{i=0}^{2} q_i F_i(p_0, p_1, p_2) = 0, \qquad F_i = F_{X_i}.$$

An equivalent way of stating this condition is to use the exponential valuation ord: $F_q(C) \to \mathbb{Z}$ associated to P. We recall below the definitions of all these objects. If $C: F(x_0, x_1, x_2) = 0$ is an irreducible projective plane curve defined by a homogeneous polynomial F of degree d and coefficients in F_q, then its homogeneous coordinate ring is

$$R_C^h = F_q[x_0, x_1, x_2]/(F),$$

where (F) is the homogeneous ideal generated by (F). The homogeneous function field is the field of quotients of R_C^h.

The elements in R_C^h determine functions on C only when they have a representation as a quotient of two forms in R_C^h of the same degree. The function field $F_q(C)$ of C is the set of all elements in the field of quotients of R_C^h which are representable in the form u/v with $u, v \in R_C^h$ and $\deg(u) = \deg(v)$. A function $z = u/v \in F_q(C)$ is defined at P if there is such a representation with $v(P) \neq 0$. The local ring at P is

$$R_P = \{z = u/v \in F_q(C): u, v \in R_C^h, v(P) \neq 0\};$$

the maximal ideal of R_P is

$$M_P = \{z = u/v \in R_P: u(P) = 0, v(P) \neq 0\}.$$

Since we are assuming that P is a simple point, we know that R_P is a unique factorization domain (Shafarevitch [76], p. 94). We introduce a local uniformizing parameter at P by means of the linear form $T = l_0 x_0 + l_1 x_1 + l_2 x_2$ in R_P which we assume crosses the curve C transversally at P, i.e. the intersection cycle $T \cdot C$ contains P with multiplicity equal to 1. Now the valuation ord_P at P of a function $f \neq 0$ is then defined as $v =: \text{ord}_P(f)$, where v is the smallest integer such that $F \in (T^v)$. To say that the point Q lies on the tangent to C at P is equivalent to the statement

$$\text{ord}_P\left(\sum_{i=0}^{2} q_i F_i(x_0, x_1, x_2) \right) \geq 1;$$

it should be remarked that the above criterion is independent of the projective coordinate system chosen.

Theorem A.4 *Let P be a simple point on the projective plane curve C rational over F_q. Of all the lines L which pass through P, there is a unique one L_0 such that $\text{ord}_P(L_0) > \text{ord}_P(L)$.*

Proof. With notations as above, we represent the coordinates of a point in the neighborhood of P by

$$X_i = \sum_{k=0}^{\infty} p_i(k) T^k, \qquad i = 0, 1, 2,$$

where $p_i(k) \in F_q$ and T is the local uniformizer chosen above. If we suppose that the line $L = A_0 X_0 + A_1 X_1 + A_2 X_2 = 0$ passes through P, then

$$L = \sum_{k=1}^{\infty} (A_0 p_0(k) + A_1 p_1(k) + A_2 p_2(k)) T^k.$$

Letting v be the least positive integer for which at least one of coefficients $p_0(v)$, $p_1(v)$, $p_2(v)$ is not zero, then $\text{ord}_P(L) = v$ if and only if $A_0 p_0(v) + A_1 p_1(v) + A_2 p_2(v) \neq 0$. Now the condition $L_0 = A_0 p_0(v) + A_1 p_1(v) + A_2 p_2(v) = 0$ determines a unique line which satisfies $\text{ord}_P(L_0) > \min_L \text{ord}_P(L)$.

Definition A.5 *The positive integer $r = \min_L \text{ord}_P(L)$, where L runs over the lines passing through P is called the order of the closed point P. A closed point is said to be linear if $r = 1$. The unique line L_0 with $\text{ord}_P(L_0) > r$ is called the tangent to the curve at P. The positive integer $s = \text{ord}_P(L_0) - r$ is the class of P.*

Theorem A.5 *If P is a point common to the curves C and C' with multiplicities m and m', then the number of intersections of C and C' at P is $\geq mm'$ with equality holding when no tangent to C at P is also a tangent to C' at P.*

Let P be a point of C: $F(x_0, x_1, x_2) = 0$; let ord_P be the valuation of $F_q(C)$. Choose a projective coordinate system so that $P = (1, 0, 0)$ and a local uniformizing parameter T for ord_P so that in a neighborhood of p the points of C are parametrized by

$$X_0 = t, \qquad X_1 = t^r, \qquad X_2 = t^{r+s} + \cdots, \qquad (A.2)$$

where $r, s > 1$, unless P is a place of a linear component of F. We exclude this case by requiring that F has no linear components, r and s are respectively the order and the class of P and $X_2 = 0$ is the tangent at P.

By Euler's theorem,

$$\sum_{i=0}^{2} x_i F_i(x_0, x_1, x_2) = nF(x_0, x_1, x_2) = 0.$$

Differentiating $F(x_0, x_1, x_2) = 0$ with respect to t we get

$$\sum_{i=0}^{2} x_i' F_i(x_0, x_1, x_2) = 0$$

Substituting the values of x_i from (A.2) into these equations we obtain

$$F_0(x_0, x_1, x_2) = F_2(x_0, x_1, x_2)\left(\left(\frac{s}{r}\right)t^{s+r} + \cdots\right),$$

$$F_1(x_0, x_1, x_2) = F_2(x_0, x_1, x_2)\left(-\left(1 + \frac{s}{r}\right)t^s + \cdots\right).$$

Designating $\text{ord}_P \, F_2(\bar{x})$ by $\delta(P)$, we have

$$\text{ord}_P\left(\sum_{i=0}^{2} q_i F_i\right) = \delta(P) + \varepsilon(P)$$

where

$$\varepsilon(P) = \begin{cases} 0 & \text{if } Q(q_0, q_1, q_2) \text{ is not on the tangent at } P \\ s & \text{if } Q \text{ is on the tangent but } Q \neq P \\ s + r & \text{if } Q = P. \end{cases}$$

Definition A.6 $\varepsilon(P)$ = *number of tangents, or the multiplicity of the tangent from Q to P.*

The total number m of tangents from F to F is

$$\sum_P \operatorname{ord}_P\left(\sum_{i=0}^{2} q_i F_i\right) = d(d-1)$$

and hence

$$\sum_P \delta(P) + \sum_P \varepsilon(P) = d(d-1)$$

or equivalently, the class of F is

$$m = d(d-1) - \sum_P \delta(P).$$

A.4 Quadratic transformations

Let $\mathbb{P}^2$ be a projective plane and denote by $(x) = (x_0, x_1, x_2)$ the coordinates of a point in some projective coordinate system: we need to consider another copy of $\mathbb{P}^2$ and denote the coordinates of a point by $(y) = (y_0, y_1, y_2)$.

Definition A.7 *A quadratic transformation is a mapping defined by*

$$T: \mathbb{P}^2 \to \mathbb{P}^2$$
$$y_0 = x_1 x_2, \qquad y_1 = x_0 x_2, \qquad y_2 = x_0 x_1. \tag{A.3}$$

The transformation T is not defined at the points

$$P_0 = (1,0,0), \qquad P_1 = (0,1,0), \qquad P_2 = (0,0,1);$$

these are called the *fundamental points* of T. These points can be thought of as the vertices of a triangle (see Figure A.1). The lines joining the fundamental points are called the *exceptional lines*. The transformation T has the following properties.

(i) T is well defined on $\mathbb{P}^2 - \{P_0, P_1, P_2\}$.
(ii) Any non-fundamental point on the line $x_i = 0$ is transformed into the point $Q_i = (\delta_{0i}, \delta_{1i}, \delta_{2i})$ where δ_{ki} is the Kronecker delta function. The mapping

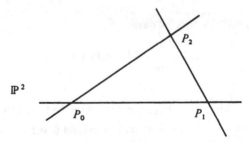

Figure A.1. $\Delta = \{x_0, x_1, x_2\} \in P^2: x_0 x_1 x_2 = 0\}$.

$$T': \mathbb{P}^2 \to \mathbb{P}^2$$

defined by

$$x_0 = y_1 y_2, \qquad x_1 = y_0 y_2, \qquad x_2 = y_0 y_1,$$

has the same properties as T above. We also have:

(iii) Any point P which does not lie on an exceptional line has an image $Q = T(P)$ which does not lie on an exceptional line of T' and $P = T'(Q)$.

If we denote by $\Delta = \{(x_0, x_1, x_2) \in \mathbb{P}^2 : x_0 x_1 x_2 = 0\}$, $\Delta' = \{(y_0, y_1, y_2) \in \mathbb{P}^2 : y_0 y_1 y_2 = 0\}$, then the mapping $T : \mathbb{P}^2 - \Delta \to \mathbb{P}^2 - \Delta'$ is biregular, i.e. it sets up a one-to-one correspondence and its inverse is given by T'.

A.4.1 Quadratic transform of a plane curve

The points (y_0, y_1, y_2) on the transform of a curve $C: F(x_0, x_1, x_2) = 0$ will satisfy the equation

$$G(y_0, y_1, y_2) = F(y_1 y_2, y_0 y_2, y_0 y_1) = 0.$$

The polynomial G is called the algebraic transform of F. To understand the geometric relation between F and G we consider the situation where F may contain one of the fundamental points of T. We first examine some simple cases. Suppose then that F is just a line:

$$L = a_0 x_0 + a_1 x_1 + a_2 a_2 = 0.$$

Three possibilities arise.

Case 1 L does not contain a fundamental point. In this case $a_0 a_1 a_2 \neq 0$ and the algebraic transform is the conic

$$C = a_0 y_2 y_1 + a_1 y_0 y_2 + a_2 y_0 y_1 = 0.$$

This conic contains each of the fundamental points of the transformation T' and no other point on an exceptional line. The fundamental points Q_0, Q_1, Q_2 of T' correspond to the points of intersection S_0, S_1, S_2 of L with the three exceptional lines of T; clearly T sets a bijective correspondence between $L - \{S_0, S_1, S_2\}$ and $C - \{Q_0, Q_1, Q_2\}$ (see Figure A.2).

Case 2 Let $L = x_1 + \lambda x_2$ be a line passing through the fundamental point $P_0 = (1, 0, 0)$. Its algebraic transform is a conic $y_0 y_2 + \lambda y_0 y_1 = 0$ consisting of two components: the exceptional line $y_0 = 0$ and the line $L' = y_2 + \lambda y_1$. As in case 1, the correspondence between L and L' is one to one. This suggests that we consider L' as the true geometric transform of L and the exceptional line $y_0 = 0$ as an inessential factor introduced by the algebra.

Case 3 The line L passes through two fundamental points, say $L = x_0 = 0$ in which case its algebraic transform $y_1 y_2 = 0$ is the union of two exceptional lines. Every point of L where T is defined goes to the point $(1, 0, 0)$ and hence we regard the transform of L to be this single point. To avoid this blow down of a line to a point

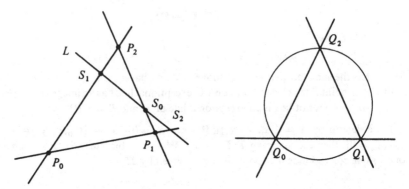

Figure A.2

we exclude from the following discussion curves with exceptional lines. The above special cases suggest the following definition.

Definition A.8 *Let* $C: F(x_0, x_1, x_2) = 0$ *be a projective plane curve with no exceptional lines as components. Let*

$$G(y_0, y_1, y_2) = F(y_1 y_2, y_0 y_2, y_0 y_1)$$

be the algebraic transform of F. *If* $G(y) = m(y)F'(y)$, *where* $m(y)$ *is a product of powers of* y_0, y_1, y_2, *and* F' *is not divisible by any* y_i, *we say that* F' *is the transform of* F *by* T.

Theorem A.6 *If* F' *is the transform of* F *under the quadratic mapping* T, *then* F *is the transform of* F' *under the inverse mapping. Furthermore, with a finite number of exceptions, the points of* F *and* F' *are in one-to-one correspondence and the components of* F *and* F' *also correspond.*

Proof. By definition we have

$$F(y_1 y_2, y_0 y_2, y_0 y_1) = M_1(y)F'(y_0, y_1 y_2); \tag{A.4}$$

and similarly

$$F'(x_1 x_2, x_0 x_2, x_0 x_1) = M_2(x)F''(x_0, x_1, x_2),$$

where F'' is the transform of F' by T'. Substituting $y_i = x_j x_k$ in (A.4) we obtain

$$F(x_0^2 x_1 x_2, x_0 x_1^2 x_2, x_0 x_1 x_2^2) = M_3(x)F'(x_1 x_2, x_0 x_2, x_0 x_1)$$
$$= M_4(x)F''(x_0, x_1, x_2),$$

where the M's are monomials in the respective variables. Since F is homogeneous, we have

$$F(x_0^2 x_1 x_2, x_0 x_1^2 x_2, x_0 x_1 x_2^2) = (x_0 x_1 x_2)^n F(x_0, x_1, x_2).$$

Since neither F nor F'' is divisible by any x_i we have $F = F''$. Also, F and F'

have only a finite number of points in common with the exceptional lines, the one-to-one correspondence between the components of F and F' follows from the formulas defining T and T'.

A.4.2 Quadratic transform of a singularity

As in case 2 above where we saw that the transform of $L = x_1 + \lambda x_2$ has two components, $y_0 = 0$ and $L' = y_2 + \lambda y_1 = 0$, we observe that the intersection of $L' = y_2 + \lambda y_1 = 0$ with the exceptional line $y_0 = 0$ at $P = (0, 1, -\lambda)$ corresponds to the direction of the line $L = x_1 + x_2$ at the fundamental point P_0. Hence there is a sense in which the directions at the point P_0 correspond to the points on the irregular line $y_0 = 0$.

To clarify this observation we show that the intersections of F' with an exceptional line correspond to the tangents to F at the corresponding fundamental point (see Figure A.3)

Theorem A.7 *Let* C: $F(x_0, x_1, x_2) = 0$ *be a projective plane curve of degree d. Suppose the fundamental points P_0, P_1, P_2 lie on C with respective multiplicities m_0, m_1, m_2 and that no exceptional line is tangent to C at any of these points. Then*

(i) *The algebraic transform G of F has the line $y_i = 0$ as a component of multiplicity m_i and the degree of the transform F' is $2d - m_0 - m_1 - m_2$.*

(ii) *There is a one-to-one correspondence preserving multiplicities between the tangents to F at P_i and the non-fundamental intersections of F' with the exceptional line $y_i = 0$.*

(iii) *F' has multiplicity $d - m_i - m_j$ at Q_k, with tangents distinct from the exceptional lines and corresponding to the non-fundamental intersections of F with $y_k = 0$.*

Proof. We only consider the point $P_0 = (1, 0, 0)$, the argument at P_1 and P_2 being similar. The homogeneous polynomial F of degree d has the form

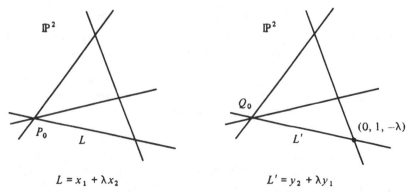

$$L = x_1 + \lambda x_2 \qquad\qquad L' = y_2 + \lambda y_1$$

Figure A.3. Directions at P_0 correspond to points on $Y_0 = 0$.

$$F(x_0, x_1, x_2) = \sum_{j=m_0}^{d} x_0^{d-j} F_j(x_1, x_2),$$

where F_j is a homogeneous polynomial of degree j in x_1, x_2 and F_{m_0}, F_d are both non-zero. The algebraic transform has the form

$$
\begin{aligned}
G(y_0, y_1, y_2) &= F(y_1 y_2, y_0 y_2, y_0 y_1) \\
&= \sum_{j=m_0}^{d} (y_1 y_2)^{d-j} F_j(y_0 y_2, y_0 y_1) \\
&= y_1^{d-m_0} y_2^{d-m_0} y_0^{m_0} F_{m_0}(y_2, y_1) + \cdots + y_0^d F_d(y_2, y_1).
\end{aligned}
$$

By assumption, $y_0^{m_0}$ is the highest power of y_0 that can be factored in G; (i) follows from this since y_1 and y_2 can treated similarly. Hence $G = y_0^{m_0} y_1^{m_1} y_2^{m_2} F'$ and deg $F' = 2d - m_0 - m_1 - m_2$. As for (ii), we need only observe that the tangents to $C: F = 0$ at $P_0 = (1,0,0)$ correspond to the linear factors in

$$F_{m_0}(x_1, x_2) = \prod_{1 \le k \le m_0} (\mu_k x_1 + \nu_k x_2),$$

while the intersections of F' with the line $y_0 = 0$ are the roots of

$$y_1^{d-m_0-m_1} y_2^{d-m_0-m_2} F_{m_0}(y_2, y_1) = 0.$$

The claim is then clear. To establish (iii) we observe that

$$F' = y_1^{d-m_0-m_1} y_2^{d-m_0-m_2} F_{m_0}(y_2, y_1) + \cdots + y_0^{d-m_0} y_1^{-m_1} y^{-m_2} F_d(y_2, y_1)$$

and the multiplicity of F' at $Q_0 = (1,0,0)$ is the degree of the polynomial

$$B(y_1, y_2) = y_1^{-m_1} y_2^{-m_2} F_d(y_2, y_1)$$

which is $d - m_1 - m_2$. Now the tangents at Q_0 correspond to the components of the curve $B(y_1, y_2) = 0$. If one of these is exceptional, $F_d(y_1, y_2)$ would be divisible by say y_1 to a power $m > m_1$. For this to occur, F would need to have m intersections with the line $x_0 = 0$ at $(0, 1, 0)$ which would then imply that one of the intersections here is the exceptional line $x_0 = 0$, contrary to assumption. The last part of (iii) follows by applying (ii) to the transformation T'.

A.4.3 Singularities off the exceptional lines

The following result guarantees that the complexity of a singularity which does not lie on an exceptional line does not increase under a quadratic transformation.

Theorem A.8 *Under a quadratic transformation, any point P of multiplicity m on the curve $C: F = 0$ which does not lie on an exceptional line goes into a point P' on the curve $C': F' = 0$ with the same multiplicity and with the tangents at P and P' having the same corresponding multiplicities.*

Proof. We may assume that $P = (1,1,1)$ so that its image under T is $P' = T(P) = (1,1,1)$; in fact if $P = (a,b,c)$ then the mapping $A: \mathbb{P}^2 \to \mathbb{P}^2$ given by $A(x_0, x_1, x_2) = (x_0', x_1', x_2')$, with $x_0' = x_0/a$, $x_1' = x_1/b$, $x_2' = x_2/c$, and $B: \mathbb{P}^2 \to \mathbb{P}^2$ given by $B(y_0, y_1, y_2) = (y_0', y_1', y_2')$, with $y_0' = y_0 a$, $y_1' = y_1 b$, $y_2' = y_2 c$, the point P becomes

$(1, 1, 1)$ and the quadratic transformation is given by $y'_0 = x'_1 x'_2$, $y'_1 = x'_0 x'_2$, $y'_2 = y'_0 y'_1$. Since the algebraic transform G and the essential factor F' differ only by components which do not contain the point P', it suffices to verify the assertion for G instead of F'. We introduce in $\mathbb{P}^2$ a new coordinate system

$$z_0 = x_0, \quad z_1 = x_1 - x_0, \quad z_2 = x_2 - x_0$$

so that the point P now has z-coordinates $(1, 0, 0)$ and

$$F(x_0, x_1, x_2) = F_1(z_0, z_1, z_2)$$

$$= z_0^{d-m} F_m(z_1, z_2) + \cdots + F_d(z_1, z_2)$$

$$= z_0^{d-m} F_m(x_1 - x_0, x_2 - x_0) + \cdots + F_d(x_1 - x_0, x_2 - x_0).$$

Hence

$$G(y_0, y_1, y_2) = (y_1 y_2)^{d-m} F_m(y_0 y_2 - y_1 y_2, y_0 y_1 - y_1 y_2)$$

$$+ \cdots + F_d(y_0 y_2 - y_1 y_2 - y_0 y).$$

In P^2 we make a similar change of coordinates

$$w_0 = y_0, \quad w_1 = y_1 - y_0, \quad w_2 = y_2 - y_0$$

so that the point P' now has coordinates $(1, 0, 0)$. With this we have

$$G(y_0, y_1, y_2) = G_1(w_0, w_1, w_2)$$

$$= (w_0 - w_1)^{d-m}(w_0 - w_2)^{d-m} F_m(w_0 w_1 - w_2 w_0, w_2 w_0 - w_1 w_2)$$

$$+ (w_0 - w_1)^{d-m-1}(w_0 - w_2)^{d-m-1} F_{m+1}(w_0 w_1 - w_2 w_0, w_2 w_0 - w_1 w_2)$$

$$+ \cdots + F_d(w_0 w_1 - w_1 w_2, w_0 w_2 - w_1 w_2)$$

$$= w_0^{2d-m} F_m(w_1, w_2) + (\cdots) + w_0^{2d-m-1} F_{m+1}(w_1, w_2)$$

$$+ (\cdots) + \cdots + w_0^d F_d(w_1, w_2) + \cdots + (\cdots).$$

where $(\cdots)$ denotes terms involving powers of w_0 of degree lower than those already accounted for. The highest power of w_0 already occurs in the first term and therefore $F_m(w_1, w_2)$ determines the tangents at P'; since $F_m(z_1, z_2)$ determines the tangents at P, the theorem is proved.

Remark The above claim simply ascertains that under a quadratic transformation. The structure of a singularity off the exceptional lines does not change.

A.4.4 Reduction of singularities

The main claim about the simplification of singularities of projective plane curves is the following result.

Theorem A.9 *By a finite sequence of quadratic transformations any irreducible projective plane curve can be transformed into one having only ordinary singularities.*

Remark Recall that a point P is an ordinary singularity if its multiplicity is $m_P \geq 2$ and all its tangents are distinct.

Proof. There is nothing to prove if the curve has no non-ordinary singularities. Suppose then that C has non-ordinary singularities of order m_P. We define the index of C to be the integer

$$I(C) = \sum_P \sum_T (m_T - 1),$$

where the sum $\sum_P$ runs over the singular points of C, and for each P, $\sum_T$ runs over the distinct tangents to C at the singular point, m_T is the multiplicity of a tangent. The aim is to show that under a suitable sequence of quadratic transformations the index of C decreases.

Let P be a non-ordinary point of F of multiplicity m. We first verify that it is possible to choose a projective coordinate system in which the curve C: $F(x_0, x_1, x_2) = 0$ has the following properties with respect to the point P:

  (i) P has coordinates $(1, 0, 0)$,
 (ii) the lines $x_1 = 0$ and $x_2 = 0$ each intersect $F(x_0, x_1, x_2) = 0$ in $d - m$ distinct points other than P;
(iii) the line $x_0 = 0$ intersects $x_1 x_2 F(x_0, x_1, x_2)$ in $n + 2$ distinct points.

The assertion (i) is clearly possible in any characteristic and can be accomplished at worst by making a finite field extension of the field of constants. As for (ii), in characteristic 0, the lines x_1 and x_2 can be obtained by choosing two distinct values of λ which make the lines $L = x_1 - \lambda x_2 = 0$ intersect $F(x_0, x_1, x_2) = 0$ in $d - m$ distinct points. That is, we take one of the derivatives of order $m - 1$ which is not identically 0 at $(1, 0, 0)$; that derivative H is a homogeneous polynomial of degree $d - m$ and upon elimination of say x_1 from H, we get a homogeneous polynomial in x_0 and x_2 whose coefficients depend on λ. The discriminant of this polynomial is itself a polynomial which is not identically zero and has at most a finite number of zeros. Taking two values λ and λ' so that $H(x_0, x_2)$ is a product of $d - m$ distinct linear factors, we obtain two lines $L = x_1 - \lambda x_2$ and $L' = x_1 - \lambda' x_2$ which can be used as $x_1 = 0$, $x_2 = 0$. In characteristic 0, (iii) is also established by a similar argument.

In characteristic $p > 0$, the situation may be far more complicated. Let us consider for example the curve

$$C: F = x_0^{p+1} - x_1^p x_2 = 0$$

over the field $\mathbb{F}_{p^2}$. The point $P = (0, 1, 0)$ is a simple point and any line passing through P and rational over F_{p^2} can be represented in the form

$$L = x_0 - \lambda x_2 = 0, \qquad \lambda \in \mathbb{F}_{p^2}.$$

Since the order of $\mathbb{F}_{p^2}^\times$ is $p^2 - 1 = (p + 1)(p - 1)$, there are λ's with $\lambda^{p+1} = 1$. Now, eliminating x_0 in F we obtain

$$F(x_2, x_1, x_2) = x_2(\lambda^{p+1} x_2^p - x_1^p) = x_2(x_2 - x_1)^p;$$

hence any such line intersects C in less than p distinct point other than $P = (0, 1, 0)$. The same is also true if we allow extensions of F_p. To remedy this situation the notion of terrible point is introduced (Fulton [26], p. 220; Hartshorne [34], p. 316).

Definition A.9 *A point P of multiplicity m on the curve C: F = 0 of degree d is called terrible if there are an infinite number of lines L through P which intersect F in fewer than d − m distinct points other than P.*

A simple argument shows that a given F can have at most one terrible point. Recall that the family of lines in $\mathbb{P}^2$ passing through a fixed point P are parametrized by the points on a projective line $\mathbb{P}^1$. If P is terrible for F, the dual curve to F contains an infinite number of points in common with a straight line. Since the dual curve is irreducible, it must be a line. In particular there are lines which intersect F in d distinct points. Observe that the property of being terrible is closely associated with the phenomenon of inseparability, in particular if p does not divide $d - m$, then the point P cannot be terrible.

If P happens to be a terrible point of multiplicity m for F and the quadratic transformation T, we first make a quadratic transformation centered at some other point Q in $\mathbb{P}^2$ of multiplicity $n = 0$ or 1 and with P not on an exceptional line. Let F' be the quadratic transform of F. By Theorem A.7 $d' = \deg F' = 2d - n$. Since $d' - m \equiv d - n \bmod p$, one of the choices $n = 0, 1$ will ensure that p does not divide $d' - m$. Then the point P' on F' corresponding to P on F will not be terrible for F'. We can thus select a projective coordinate system with $x_1' = 0$, $x_2' = 0$ each intersecting $F' = 0$ in $d - m$ distinct points other than P and for the line $x_0' = 0$ we take a line which intersects $x_1 x_2 F(x_0, x_1, x_2) = 0$ in $d + 2$ distinct points.

Once the conditions (i)–(iii) are satisfied with respect to P, we can apply Theorem A.7 with $m_0 = m$, $m_1 = m_2 = 0$. We thus have a transformation

$$T: \mathbb{P}^2 \to \mathbb{P}^2$$

$$C: F(x_0, x_1, x_2) = 0 \to C^1: F^1(y_0, y_1, y_2) = 0$$

with the point P being mapped to the distinct points $\{P_1', \ldots, P_k'\}$. Furthermore C^1 now satisfies the following additional properties:

(iv) $\deg F^1 = 2d - m$.
(v) any singularity F other than P is transformed into one of the same multiplicity, with ordinary singularities going into ordinary singularities;
(vi) F^1 has three new singularities, one of order d and two of order $d - m$;
(vii) corresponding to P, F^1 has certain points $P_1', \ldots, P_k'$ on the exceptional line $y_0 = 0$ with multiplicities $m_1', \ldots, m_k'$.

Remark We observe that under a quadratic transformation, the old singularities at the fundamental points are separated, i.e. branches centered at the fundamental points are separated. At the fundamental points in the new $\mathbb{P}^2$, ordinary singularities are introduced which do not affect the further discussion of non-ordinary singularities.

The above remark implies that the index of a curve under a quadratic transformation can only decrease or remain the same.

Recall that the deficiency of a projective plane curve of degree d with multiple points $P_0, P_1, \ldots, P_t$ of multiplicity $m_0, m_1, \ldots, m_t$ is defined by

$$g^*(C) = \frac{1}{2}(d-1)(d-2) - \frac{1}{2}\sum_{i=0}^{t} m_i(m_i - 1).$$

If we let $C': F'(x_0, x_1, x_2) = 0$ be the curve satisfying (iv)–(vii) above we get, by enumerating the new singularities and using Theorem A.7,

$$g^*(C') = g^*(C) - \frac{1}{2}\sum_{j=1}^{t} m'_j(m'_j - 1).$$

The direction which the argument must now take is clear: we choose quadratic transformations centered at singular points which contribute to the index. When the index $I(C)$ does not change, the point P_0 has been mapped to a single point P'_1, where the multiplicity of the tangent remains the same; when this happens the deficiency $g^*(C)$ decreases, i.e.

$$g^*(C') = g^*(C) - \tfrac{1}{2}m'_1(m'_1 - 1) < g^*(C).$$

After getting rid of a terrible point if necessary, we can perform a new quadratic transformation centered at P'_1. Then again either the index has diminished, in which case one of the non-ordinary singularities has become simpler, or the deficiency has decreased. The crucial observation is that the index must eventually decrease because the deficiency cannot diminish forever by Theorem A.3. This concludes the proof of the theorem.

Bibliography

[1] Abhyankar, S. S., Historical ramblings in algebraic geometry and related algebra, *A. M. S. Monthly*, **83** (1976), 409–48.

[2] Atiyah, M. F. & Macdonald, I. G., *Introduction to Commutative Algebra*, Addison-Wesley, Reading, Mass., (1969).

[3] Atkin, A. O. L., Modular forms of weight one, and supersingular equations (preprint, University of Illinois, Chicago, 1975).

[4] Artin, E., Quadratische Korper im Gebiet der hoheren Kongruenze I, II, *Math. Zeit.*, **19** (1924), 153–246.

[5] Berlekamp, E., *Algebraic Coding Theory*, McGraw-Hill, New York, 1968.

[6] Berndt, B. & Evans, R., Gauss sums, *Bull. A. M. S.*, **5** (1981), 107–29.

[7] Bombieri, E., On exponential sums in finite fields, *Am. J. of Math.*, **88** (1966), 71–105.

[8] Bombieri, E., Counting points on curves over finite fields (d'après S. A. Stepanov), *Sém. Bourbaki*, No. 430 (1972/3).

[9] Bombieri, E., Hilbert's 8th problem: an analogue, *Proc. Symposia in Pure Math. (A. M. S.)*, **28** (1976), 269–74.

[10] Carlitz, L. & Uchiyama, S., Bounds for exponential sums, *Duke Math. J.*, **24** (1957), 37–41.

[11] Chevalley, C., *Introduction to the Theory of Functions of One Variable* A. M. S. Math. Surveys, New York, 1951.

[12] Deligne, P. & Rappoport, M., Schémas des modules des courbes elliptiques, *Lecture Notes in Math.*, **349** (1973), 143–316.

[13] Deligne, P., La conjecture de Weil, I, *Publ. Math. IHES*, **43** (1974), 273–308.

[14] Deligne, P., Applications de la formule des traces aux sommes trigonometriques, SGA4.5, *Lecture Notes in Math.*, **569** (1977), 168–232.

[15] Deuring, M., Die Typen der Multiplikatorenringe elliptischer Funktionenkorper, *Abh. Math. Sem. Univ. Hamburg*, **14** (1941), 197–272.

239

[16] Deuring, M., Lectures on the Theory of Algebraic Functions of one variable, *Lecture Notes in Math.*, **314** (1973).

[17] Drinfeld, V. G., Elliptic modules, *Mat. Sb.*, **94** (1974), 594–627.

[18] Drinfeld, V. G. & Vladut, S. G., Number of points of an algebraic curve, *Functional Anal. and Appl.*, **17** (1983), 68–69.

[19] Driencourt, Y., Some properties of elliptic codes over a field of characteristic 2, *Springer Lecture Notes in Comp. Sci.*, vol. 229, 1985.

[20] Driencourt, Y. & Michon, J. F., Remarques sur les codes géometriques, *C. R. Acad. Sci. Paris*, **301** (1985), 15–17.

[21] Davenport H. & Hasse, H., Die Nullstellen der Kongruenzzetafunktion in gewissen zyklischen Fallen, *Crelle*, **172** (1935), 151–82.

[22] Davenport, J. H., On the integration of algebraic functions, *Springer Lecture Notes in Comp. Sci.*, vol. 102, 1981.

[23] Eichler, M., Quaternary quadratische Formen und die Riemannsche Vermutung fur die Kongruenzzetafunktion, *Arch. Math.*, **5** (1954), 355–66.

[24] Eichler, M., *Introduction to the Theory of Algebraic Numbers and Functions*, Academic Press, New York, 1966.

[25] Evans, R., Identities for products of Gauss sums, *L'Enseignement mathématique*, **27** (1981), 197–209.

[26] Fulton, W., *Algebraic Curves*, Benjamin, New York, 1969.

[27] Goldfeld, D., and Sarnak, P., Sums of Kloosterman sums, *Inventiones Math.*, **71** (1983), 243–50.

[28] Goppa, V. D., Decoding and diophantine approximations, *Problems of control and information theory*, **5** (1976), 1–12.

[29] Goppa, V. D., Codes on algebraic curves, *Soviet Math. Dokl.*, **24** (1981), 170–2.

[30] Goppa, V. D., Algebraico-geometric codes, *Math. USSR Izvestiya*, **21** (1983), 75–91.

[31] Goppa, V. D., Coding and information theory, *Russ. Math. Surveys*, **39** (1985), 87–141.

[32] Gross, B. H., and Koblitz, N., Gauss sums and the p-adic gamma function *Ann. of Math.*, **109** (1979), 569–581.

[33] Hansen, J. P., Codes on the Klein quartic, ideals and decoding, *IEEE Trans. Info. Theory*, **IT-33** (1987), 923–5.

[34] Hartshorne, R., *Algebraic Geometry*, Springer, New York, 1977.

[35] Hijikata, H., Explicit formula for the traces of the Hecke operators for $I(N)$, *J. Math. Soc. Japan*, **26** (1974), 56–80.

[36] Hirschfeld, J. W. P., Linear codes and algebraic curves, in *Geometrical Combinatorics*, eds. F. C. Holroyd and R. J. Wilson, Pitman, Boston, 1984.

[37] Igusa, J., Class number of a definite quaternion with prime discriminant, *Proc. N. Acad. Sci.*, **44** (1958), 312–14.

[38] Igusa, J., Kroneckerian model of fields of elliptic modular functions, *Amer. J. Math.*, **81** (1959), 561–77.

[39] Igusa, J., On the transformation theory of elliptic functions, *Amer. J. Math.*, **81** (1959), 436–52.

[40] Ihara, Y., Some remarks on the number of rational points of algebraic curves over finite fields, *J. Fac. Sci., Univ. Tokyo*, **28** (1981), 721–4.

[41] Katsman, G. L., Tsfasman, A. M., and Vladut, S. G., Modular curves and codes with polynomial construction, *IEEE Trans. Information Theory*, **30** (1984), 353–5.

[42] Katsman, G. L., Tsfasman, A. M., and Vladut, S. G., Modular curves and codes with polynomial complexity, *Problemy Pederachi Informatsii* **20** (1984), 47–55.

[43] Katz, N., *P*-adic properties of modular schemes and modular forms, Proc. Antwerp Conf., *Lecture Notes in Math.*, **350** (1972), 69–190.

[44] Katz, N., Sommes Exponentielles, *Astérisque*, **79** (1980).

[45] Katz, N., and Mazur, B., Elliptic curves and modular curves, *Ann. of Math. Studies*, Princeton, 1985.

[46] Katz, N., Gauss sums, Kloosterman sums, and Monodromy groups, *Ann. of Math. Studies*, Princeton, 1988.

[47] Lachaud, G., Les codes géometriques de Goppa, *Sem. Bourbaki*, no. 641, 1985.

[48] Lachaud, G. & Wolfmann, J., Sommes de Kloosterman, courves elliptiques et codes cycliques en caracteristique 2, *C. R. Acad. Sci. Paris*, **305** (1987), Sér. I, 881–3.

[49] Leidel, J., and Niederreiter, H., Finite fields, Addison-Wesley, New York, 1983.

[50] Lefschetz, S., The early development of algebraic geometry, *A. M. S. Monthly*, **76** (1969), 451–60.

[51] Livne, R., Coding theory and Kloosterman sums (Lecture at Baruch College, New York, September 1988).

[52] Loo-Keng, H., On exponential sums, *Science Record*, **1** (1957), 1–4.

[53] Lusztig, G., Representations of finite Chevalley groups, *A. M. S. regional conference*, vol. 39, 1978.

[54] Manin, Y. I., What is the maximum number of points on a curve over *F Jour. Fac. Sci., Univ. Tokyo*, **28** (1981), 715–20.

[55] Manin, Y. I. and Vladut, S. G., Linear codes and modular curves, *Jour. Soviet Math.*, **30** (1985), 2611–43.

[56] McEliece, R. J., Rumsey, H., Euler products, cyclotomy, and coding *J. Number Theory*, **4** (1972), 302–11.

[57] McEliece, R. J., *The Theory of Information and Coding*, Addison-Wesley, Reading, Mass., 1982.

[58] Mazur, B., Modular curves and the Eisenstein ideal, *Publ. Math. IHES*, **50** (1978), 34–186.

[59] Mazur, B., Eigenvalues of Frobenius acting on algebraic varieties over finite fields, *Proc. Symposia Pure Math.*, **29** (1975), 231–61.

[60] Mazur, B., Arithmetic on curves, A. M. S. Colloquium Lectures, *Bull. A. M. S.*, **14**, No. 2 (1986), 206–59.

[61] Ogg, A. *Modular forms and Dirichlet Series*, Benjamin, New York, 1969.

[62] Ogg, A., On the reduction modulo p of $X_0(pM)$ (preprint, 1976).

[63] Perelmuter, G. I., On certain character sums, *Uspehi Mat. Nauk*, vol. **18** (1963), 145–9.

[64] Perelmuter, G. I., Estimation of a sum along an algebraic curve, *Mat. Zametki*, **5** (1969), 373–80.

[65] Porter, S. C., Decoding codes arising from goppa's construction on algebraic curves, Ph.D. Thesis, Yale Univ., 1988.

[66] Robert, A., Elliptic Curves, *Lecture Notes in Math.*, vol. 326, Springer, New York, 1973.

[67] Samuel, P., Lectures on old and new results on algebraic curves, *Tata Institute Lectures on Math.*, 36, Bombay, 1966.

[68] Schmid, H. L., and Teichmuller, O., Ein neuer Beweis fur die Funtionalgleichung der L-Reihen, *Abh. Math. Sem. Univ. Hanburg*, **15** (1943), 85–96.

[69] Schmidt, W. M., Equations Over Finite Fields: An Elementary Approach, *Lecture Notes in Math.*, vol. 536, Springer, New York, 1976.

[70] Serre, J.-P., *Groupes Algébriques et Corps de Classes*, Herman, Paris (1959).

[71] Serre, J.-P., Zeta and L-functions, *Arithmetical Algebraic Geometry*, Harper & Row, New York, 1965, 82–92.

[72] Serre, J.-P., *Corps Locaux*, Herman, Paris (1968).

[73] Serre, J.-P., Majorations de sommes exponentielles, *Astérisque*, **41–2** (1977), 111–26.

[74] Serre, J.-P., Sur le nombre des points rationnels d'une courbe algebrique sur un corps fini, *C. R. Acad. Sci. Paris*, **297** (1983), Sér. I: 397–401.

[75] Serre, J.-P., Sur les représentations modulaires de degré 2 de $\mathrm{Gal}(\bar{Q}/Q)$, *Duke Math. Jour.*, 54, No. 2 (1987).

[76] Shafarevich, I. R., *Basic Algebraic Geometry*, Springer, New York, 1974.

[77] Shannon, C. E., A Mathematical Theory of Communication, *Bell System Tech. J.*, **27** (1948), 379–423, 623–56.

[78] Shimura, G., *Introduction to the Arithmetic Theory of Automorphic Forms*, Publ. Math. Soc. Japan, vol. 11 (1971).

[79] Shimura, G., Correspondances modulaires et les fonctions zéta de courbes algébriques, *J. Math. Soc. Japan*, **10** (1958), 1–28.

[80] Silverman, J. H., The arithmetic of elliptic curves, *Graduate Texts in Math.*, vol. 106, Springer, New York, 1986.

[81] Sloane, N. J. A., Error-correcting codes and invariant theory: new applications of a nineteenth-century technique, *A. M. M. Monthly*, **84** (1977), 82–107.

[82] Smith, R. A., On n-dimensional Kloosterman sums, *C. R. Math. Rep. Acad. Sci. Canada*, **1** (1979), 173–6.

[83] Smith, R. A., The distribution of rational points on hypersurfaces defined over a finite field, *Mathematika*, **17** (1970), 328–32.

[84] Stark, H. M., On the Riemann Hypothesis in hyperelliptic function fields, *Proc. Symposia Pure Math.*, **24** (1973), 285–302.

[85] Stickelberger, L., Uber eine Verallgemeinerung der Kreistheilung, *Math. Ann.*, **37** (1890) 321–67.

[86] Stohr, K.-O. and Voloch, J. F., Weierstrass points and curves over finite fields, *Proc. London Math. Soc.*, vol. 52 (1986), 1–19.

[87] Swinnerton-Dyer, H. P. F., and Birch, B. J., Elliptic curves and modular functions, *Lecture Notes in Math.*, **476** (1974), 2–32.

[88] Tate, J., Residues of differentials on curves, *Ann. Scient. Ec. Norm. Sup.*, **4** (1968), 149–59.

[89] Tate, J., The arithmetic of elliptic curves, *Inventiones Math.*, **23** (1974), 179–206.

[90] Tsfasman, M. A., Vladut, S. G., and Zink, Th., Modular curves, Shimura curves, and Goppa codes, better than Varshamov–Gilbert bound, *Math. Nachr.*, **109** (1982), 21–8.

[91] van Lint, J. H., *Introduction to Coding Theory*, Springer, New York, 1982.

[92] van Lint, J. H. and Springer, T. A., Generalized Reed-Solomon codes from algebraic geometry, *IEEE Trans. Info. Theory*, **IT-33** (1987), 305–9.

[93] Vasquez, A. T., Some explicit self-dual families of error-correcting codes based on Goppa's algebraico-geometric construction (preprint, Graduate Center, C. U. N. Y., 1987).

[94] Walker, R. J., *Algebraic Curves*, Dover, New York, 1950.

[95] Weil, A., Sur les courbes algébriques et les variétés qui s'en déduisent, Herman, Paris, 1948.

[96] Weil, A., On some exponential sums, *Proc. Nat. Acad. Sci. USA*, **34** (1948), 204–7.

[97] Weil, A., Number of solutions of equations in finite fields, *Bull. A. M. S.*, **55** (1949), 497–508.

[98] Weil, A., Review of 'Introduction to the theory of algebraic functions of one variable', *Bull. A. M. S.*, **57** (1951), 384–98.

[99] Weil, A., *Basic Number Theory*, Springer, New York (1967).

[100] Weil, A., La cyclotomie jadis et naguère, *L'Enseignement Math.*, **20** (1974), 247–63.

[101] Weil, A., Zur algebraichen Theorie der algebraichen Funktionen, *Crelle*, **179** (1938), 129–33.

[102] Yui, N., Explicit form of the modular equation, *Crelle*, **299** (1978), 185–200.

Index

Index

Printed in the United States
By Bookmasters